The Consumer Guide to Microplastics

What We Know, What We Don't, and Practical Ways to Reduce Exposure

Brian D. Schell

Published by Schell-IP
First Edition

Library of Congress Control Number: 2026912447

ISBN: 979-8-9956873-4-4

Printed in the United States of America
Lebanon, Pennsylvania
2026

For information, speaking inquiries, or educational use:
bschell1959@gmail.com
www.schell-ip.com

Table of Contents

Introduction

Most people are hearing more about microplastics.

They are being found in water, food, air, household dust, and even inside the human body. The topic now appears regularly in news reports, scientific studies, government discussions, and everyday conversations.

At the same time, the information can feel difficult to sort through.

Some sources describe microplastics as a major emerging health concern. Others emphasize that many questions are still being studied and that the long-term effects are not yet fully understood. Between those two positions, many people are left trying to determine what actually matters in daily life.

This book is written to help close that gap.

It is not a technical manual, a scientific journal, or a warning designed to create fear. It is a practical guide to understanding where microplastics come from, how people encounter them in everyday life, and what can reasonably be done to reduce unnecessary exposure.

Microplastics are not a single-product problem. They are connected to packaging, textiles, transportation, water infrastructure, consumer behavior, and material design. Understanding them requires looking at how these areas interact over time.

A broader perspective matters.

Microplastics are connected to many parts of modern life, including packaging, transportation, textiles, water infrastructure, household environments, and consumer behavior.

Plastic became widely used because it solved real problems. It protects food, reduces breakage, lowers transportation costs, improves sanitation, and supports modern manufacturing and distribution. These benefits are real and important.

At the same time, the durability that made plastics successful also creates long-term challenges. Materials designed to last do not simply disappear after use. Over time, they break down into smaller and smaller fragments.

Those fragments are now being detected more often because measurement technology has improved. Scientists can identify particles today that could not easily be measured years ago.

Research into microplastics is expanding rapidly. Important work is being done in environmental science, toxicology, medicine, materials engineering, and public health. Some findings are well established. Other areas remain uncertain.

This book respects both realities.

Where scientific understanding is clear, that will be stated directly. Where uncertainty remains, that will also be stated clearly. The goal is not to overstate or dismiss the issue, but to explain it in a balanced and understandable way.

The focus throughout this book is practical:

- where microplastics come from
- how exposure happens
- which everyday situations matter most
- and what reasonable steps consumers can take without overreaction or fear

The goal is reduction and awareness, not perfection.

No modern environment is completely free from microplastics. They are now part of the world people live in. That does not mean individuals are powerless. Small, informed decisions can reduce unnecessary exposure over time.

This book is also written with students, educators, and professionals in mind. Each chapter includes discussion questions and application exercises designed to encourage practical thinking rather than simple conclusions.

Microplastics are often presented as either a scientific debate or an environmental headline.

They are also a real-world systems issue.

Understanding that may be the most important step of all.

Chapter 1: What Microplastics Are

Microplastics are small pieces of plastic.

That is the simplest and most accurate way to start.

They are not a special type of material. They are not a new invention. They are fragments of the same plastics used every day in packaging, products, textiles, and manufacturing.

The difference is size.

Microplastics are generally defined as plastic particles smaller than five millimeters. Many are much smaller. Some are small enough that they cannot be seen without magnification.

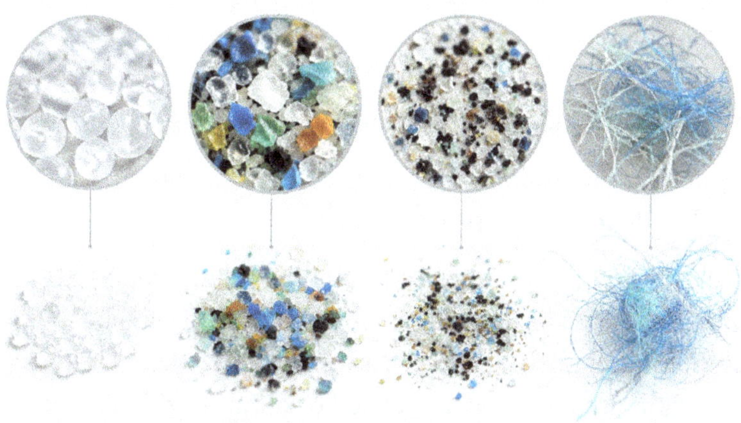

Microplastics vary widely in size, shape, and source, ranging from fragments and pellets to fibers produced through everyday material use and breakdown.

There are two main ways they form.

The first is through breakdown. Larger plastic items wear down over time. Sunlight, heat, mechanical stress, and repeated handling all contribute. A plastic bottle, a food container, or a piece of packaging does not simply disappear. It gradually breaks into smaller and smaller pieces.

Everyday Wear Creates Particles

A reusable food container may appear unchanged after years of use, but repeated opening, closing, washing, stacking, and temperature changes slowly wear the material over time.

The same principle applies to many common products:

- bottled water containers
- synthetic clothing
- packaging films
- automotive materials
- household storage containers

Most people never notice this gradual wear because the particles are extremely small. The material does not disappear all at once. It slowly fragments through normal use and handling.

This is one reason microplastics are considered a broader material issue rather than a single-product issue.

The second is through direct production. Some microplastics are intentionally created at a small size. These can be used in manufacturing, industrial processes, or specialized applications. While certain uses have been reduced or eliminated in recent years, the broader category still exists.

In everyday life, most microplastics people encounter come from breakdown.

A plastic package that is handled, opened, flexed, and exposed to temperature changes will shed small particles over time. Synthetic fabrics release fibers during wear and washing. Tire wear produces fine particles that enter air and water systems. Even routine activities like opening a container or washing clothes contribute small amounts.

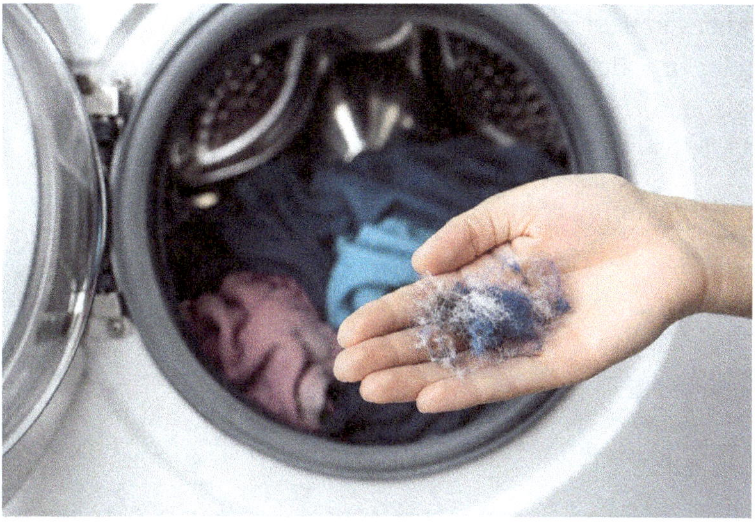

Synthetic fabrics can release small fibers during normal wear and washing, contributing to microplastic pathways in everyday life.

None of this is new behavior.

What is new is the ability to measure it.

Improved Detection Changes Visibility

Many industrial supply chains operate for years before measurement technology improves enough to detect smaller defects, particles, or contaminants.

The same pattern applies to microplastics.

As scientific tools became more sensitive, researchers began detecting particles that previously could not be measured reliably in:

- water
- food
- air
- environmental samples

In many cases, the particles were likely already present. The difference is that technology can now identify them at much smaller scales.

Advances in testing have made it possible to detect very small particles in water, food, and air. As detection improves, more is being found. That does not necessarily mean the problem is suddenly worse. It means it is now visible.

That distinction matters.

Advances in testing technology now allow researchers to detect extremely small plastic particles in water, food, air, and environmental samples.

Microplastics are part of a larger system. They are a result of how materials are designed, used, and managed over time. They reflect durability, convenience, cost, and performance decisions that have been made for decades.

Plastic works well because it is durable and stable. Those same qualities mean it does not easily break down into natural components. Instead, it fragments.

Durability Creates Tradeoffs

Manufacturers often choose plastics because they are:

- lightweight

- durable
- moisture resistant
- cost effective
- resistant to breakage

Those same qualities help reduce shipping damage, improve food protection, and lower transportation costs.

However, durability also means many plastics do not quickly return to natural components after use. Instead, they remain in the environment longer and gradually fragment over time.

The challenge is not simply the material itself. It is how modern Industries use, recover, and manage those materials across their full lifecycle.

Understanding that helps keep the issue in perspective.

Microplastics are not a single source problem. They come from many pathways. Packaging, textiles, transportation, and industrial activities all contribute. No single product or material explains the whole picture.

They are also not all the same.

Different plastics behave differently. Size, shape, and composition vary widely. Some particles are rigid. Others are fibers. Some come from clear packaging films. Others come from colored or engineered materials. These differences matter when scientists study how microplastics move through the environment and interact with living organisms.

At the same time, research is still developing.

Microplastics have been detected in water, food, air, and in human samples. What that means for long-term health is still being studied. There is active research, but there are also gaps in understanding.

This book does not attempt to resolve that uncertainty.

Instead, it focuses on what can be understood clearly today:

- where microplastics come from
- how they show up in daily life
- and what people can realistically do about them

The goal is not to simplify the issue beyond reality. It is to make it understandable without losing accuracy.

Microplastics are a byproduct of modern material systems.

To understand them, it helps to look at how those material networks actually work.

Microplastics originate from many everyday products and systems, moving through water, air, soil, and household environments over time.

Circular Exposure

Much of the modern sustainability conversation focuses on creating a "circular economy."

The idea sounds positive: materials are reused, recycled, and kept within productive cycles rather than discarded as waste.

But microplastics reveal a different kind of circular system.

Plastic particles do not simply disappear after use.

Over time, many move through interconnected environmental pathways:

- air
- water
- soil
- food systems
- indoor dust
- human activity

As larger plastics break down, smaller particles can travel through rivers, oceans, agricultural land, wastewater infrastructure, and atmospheric movement.

Some eventually return to people through:

- drinking water
- food
- airborne exposure
- everyday environmental contact

In that sense, microplastics create a form of circular exposure.

Materials designed for durability continue circulating long after their original purpose has ended.

This does not mean every exposure creates immediate harm. Research is still developing, and many scientific questions remain under study.

However, the broader pattern is becoming difficult to ignore.

Modern material networks were largely designed around performance, convenience, cost, and durability. Far less attention was historically placed on how persistent materials would behave after decades of fragmentation and environmental circulation.

Microplastics are forcing a larger question:

What happens when materials designed to last begin moving continuously through the natural systems that support human life?

Key Takeaways

- Microplastics are small pieces of common plastics, not a new material
- Most microplastics come from breakdown of larger products over time
- Everyday use, not just disposal, contributes to their formation
- Detection has improved, which is why more is being found
- Microplastics reflect how modern materials are designed and used

Discussion Questions

1. Why is it important to distinguish between detection and actual increase in microplastics?
2. Which everyday products are most likely to contribute to microplastic formation?
3. How does durability, which is normally a benefit, contribute to this issue?
4. Why is it misleading to think of microplastics as coming from a single source?

Application Exercise

Choose one product you use daily.

Identify:

- what material it is made from
- how it is handled during use
- where microplastic formation could occur over time

Keep the focus on observation, not speculation.

Chapter 2: How They Enter Everyday Life

Microplastics are often described as being "everywhere".

That can feel abstract.

A more useful way to understand the issue is to look at how they move through everyday systems. People do not encounter microplastics as an idea. They encounter them through water, food, air, clothing, and routine activities.

These are the pathways that matter.

Microplastics move through everyday systems including water, food, air, clothing, and packaging as part of normal daily life.

Water

Water is one of the most discussed sources.

Microplastics have been detected in both bottled and tap water. The sources vary. In some cases, particles come from environmental exposure. In others, they come from packaging, handling, or distribution systems.

For bottled water, the container itself is part of the system. Filling, sealing, transport, and storage all involve contact with plastic materials. Over time, small particles can be introduced.

For tap water, the system is different. Water moves through treatment facilities and distribution networks. Microplastics can enter from upstream sources or from infrastructure.

The key point is that water reflects the infrastructure, packaging, and handling involved before it reaches consumers.

Different Systems, Different Exposure

Two people may drink the same amount of water each day but encounter different exposure conditions depending on the system delivering it.

Bottled water moves through:

- filling systems
- plastic packaging
- transportation
- storage environments

Tap water moves through:

- treatment facilities
- distribution infrastructure
- household plumbing systems

Both systems are designed to deliver safe drinking water. However, each system creates different opportunities for particle introduction along the way.

This is why discussions about microplastics often focus less on a single source and more on the entire process surrounding use and delivery.

Water exposure reflects the systems, infrastructure, packaging, and handling involved before consumption.

Food and Packaging

Food is another pathway, often connected to packaging.

Plastic packaging is designed to protect products. It performs well in that role. It prevents contamination, extends shelf life, and supports distribution.

At the same time, it creates contact between food and material.

Opening, closing, flexing, and handling packaging can release small particles. Heat may increase interaction in some conditions. Processing and packaging environments also play a role.

Packaging Designed for Performance

Food packaging is engineered to solve practical problems:

- protecting products
- preventing contamination
- extending shelf life
- reducing shipping damage

Modern distribution systems depend heavily on these materials because they are lightweight, durable, and efficient.

At the same time, repeated handling and material contact become part of the overall exposure system. The same features that improve convenience and product protection can also create additional opportunities for material wear over time.

This does not mean all packaging creates the same level of exposure. Materials, design, and use conditions matter. But the connection exists, and it is part of the broader system.

Food packaging protects products and extends shelf life, but also creates repeated contact between materials and consumers during everyday use.

Air and Dust

Air is less visible but important.

Microplastics are present in indoor and outdoor air. Much of indoor exposure comes from fibers. Synthetic textiles, carpets, and upholstery shed small particles during normal use.

These particles settle as dust and can become airborne again through movement, cleaning, and airflow.

Indoor dust can contain fibers and particles released from textiles, carpets, upholstery, and other everyday household materials.

Indoor environments often have higher concentrations than outdoor spaces. People spend a significant amount of time indoors. That makes this pathway relevant, even if it receives less attention.

Clothing and Laundry

Clothing is a direct and continuous source.

Synthetic fabrics release fibers during wear. Washing increases this effect. Laundry systems can carry fibers into wastewater streams, where they may not be fully removed.

This is not limited to specialized products. Common materials used in everyday clothing contribute to the overall load.

Again, this reflects design choices. Synthetic fibers offer performance benefits. They are durable, flexible, and cost-effective. The tradeoff is fiber release over time.

Synthetic clothing materials can release small fibers during normal wear and repeated washing over time.

Everyday Handling

Small, routine actions also contribute.

Opening a container. Cutting a package. Using a plastic utensil. Handling products made with polymer materials. These are normal activities.

Each one may release very small amounts.

Individually, the impact is minor. Across millions of uses, it adds up.

Small Actions Become Large System Effects

Many large-scale industrial and environmental effects are created through repetition rather than a single event.

A single package opening or plastic utensil use may release only a very small amount of material. On its own, the effect appears insignificant.

But when repeated across millions of households and billions of daily interactions, small individual actions become measurable system-level activity.

This is one reason researchers focus on overall material use patterns rather than isolated products.

Many large-scale environmental effects emerge through repeated everyday interactions rather than a single event.

A System, not a Single Source

One of the most important points is this:

Microplastics do not come from one place.

They are the result of multiple systems operating at the same time:

- packaging systems
- textile systems
- transportation systems
- water systems
- consumer use patterns

Focusing on one source alone can be misleading. Reducing exposure is not about eliminating a single product. It is about understanding where exposure is more likely and making informed choices.

Performance Materials Create Tradeoffs

Synthetic fabrics became widely used because they offer important advantages:

- durability
- stretch and flexibility
- wrinkle resistance
- moisture management
- lower cost

These materials perform well in everyday clothing and activewear.

However, repeated wear and washing gradually release small fibers into laundry systems and indoor environments. The issue is not simply the presence of synthetic materials, but how large-scale use and repeated washing interact over time.

That requires a practical view.

Some exposure is difficult to avoid completely. That does not mean nothing can be done. It means the goal is reduction, not perfection.

In the chapters that follow, the focus will shift from where microplastics come from to what is currently known, what is still uncertain, and what actions make sense in everyday life.

Understanding the pathways is the first step.

From there, decisions become clearer.

Everyday household and urban systems can carry fibers and particles into wastewater and environmental pathways over time.

Key Takeaways

- Microplastics enter daily life through water, food, air, and routine use
- Exposure reflects systems, not a single source

- Packaging, textiles, and indoor environments are major contributors
- Small, repeated actions contribute more than isolated events
- Reducing exposure is about informed choices, not complete elimination

Discussion Questions

1. Which exposure pathway seems most significant in daily life and why?
2. How do indoor environments change the exposure picture?
3. What role does packaging play compared to other sources?
4. Why is focusing on a single source often misleading?

Application Exercise

Map your typical day from morning to evening.

List where microplastic exposure could occur:

- water
- food
- air
- product use

Then identify one realistic change you could make without disrupting your routine.

Chapter 3: Water - Tap vs Bottled

Water is one of the most discussed sources of microplastics.

It is also one of the most misunderstood.

Many people assume there is a simple answer to the question: Is bottled water safer or cleaner than tap water?

The reality is more complicated.

Both bottled and tap water can contain microplastics. The difference is how those particles enter the system and what factors influence them.

Understanding that requires looking at how each one actually works.

Bottled Water

Bottled water is a packaged product.

It is filled, sealed, transported, stored, displayed, purchased, and consumed within a packaging system designed to protect the contents and maintain consistency.

Plastic bottles are widely used because they are:

- lightweight
- durable
- resistant to breakage
- cost-effective

These are practical advantages.

At the same time, bottled water remains in continuous contact with packaging materials throughout the life of the product.

A Case of Water Stored in Summer Heat

Consider a pallet of bottled water shipped during the summer.

The bottles may begin at a filling plant, move through warehouses, travel in trailers, sit at distribution centers, and eventually remain stacked outside a retail store or inside a hot vehicle.

The bottles remain lightweight, durable, and resistant to breakage.

At the same time, the product experiences changing environmental conditions:

- vibration during transport
- pressure from stacking
- repeated handling
- temperature fluctuations
- extended storage time

In manufacturing systems, heat and friction often influence how packaging materials behave over time. Plastic packaging is no exception.

For example, bottled water left for extended periods in a hot vehicle may experience a very different storage environment than water kept refrigerated or stored indoors at stable temperatures.

This does not mean bottled water is automatically unsafe.

It means the entire packaging and distribution system influences the final product.

Understanding microplastics requires looking beyond the water itself and examining the full lifecycle of the package:

- the bottle
- the cap
- transportation
- storage conditions
- handling over time

The surrounding conditions matter as much as the product itself.

Packaging Designed for Performance

In manufacturing and distribution, packaging must survive handling, shipping, stacking, and storage without failing.

Plastic bottles perform well because they are durable and flexible. A glass bottle may break. A paper container may weaken from moisture.

Plastic solves those problems efficiently.

The same material durability that improves performance also means the packaging remains stable and persistent over time.

During filling and sealing, bottles are exposed to pressure, friction, and handling. Caps are tightened to create secure seals. Cases are stacked and transported over long distances.

Each interaction is small.

Over time, repeated handling and contact can contribute to the release of very small particles.

This does not mean every bottle creates measurable exposure. It means the system allows for interaction between packaging and contents.

Repeated opening and closing creates friction and wear at the bottle cap and thread interface over time.

Storage and Time

Storage conditions also matter.

Water may remain in a bottle for weeks or months before it is consumed. During that time, the bottle may experience:

- temperature changes
- sunlight exposure
- physical movement
- repeated handling

Heat is particularly important.

Materials often behave differently at higher temperatures. This is true across many manufacturing and packaging systems.

For example, leaving bottled water in a hot vehicle for extended periods creates a different environment than storing it in a cool indoor location.

Time also matters because prolonged contact increases the duration of interaction between water and packaging.

Tap Water

Tap water moves through a different type of system.

Instead of individual containers, tap water depends on:

- water sources
- treatment facilities
- distribution infrastructure
- household plumbing systems

Microplastics can enter at multiple stages.

Possible sources include:

- environmental contamination upstream

- particles entering rivers and reservoirs
- infrastructure wear
- limitations in filtration and treatment systems

Tap water reflects the condition of the infrastructure and treatment process behind it.

A homeowner may assume that turning on the faucet is a simple process.

In reality, tap water may have traveled many miles through a large interconnected system before reaching the home.

That system can include:

- reservoirs or rivers
- pumping stations
- treatment facilities
- underground pipes
- neighborhood distribution lines
- household plumbing

Water quality depends on the condition and performance of the entire infrastructure, not just the final delivery point.

For example, two communities may receive water from different sources and operate different treatment technologies. One system may have newer infrastructure and advanced filtration, while another may rely on older pipes and equipment installed decades earlier.

Even after treatment, water continues moving through aging infrastructure before reaching consumers.

Water treatment facilities are highly important for public health and are designed to remove many contaminants.

However, most municipal systems were developed long before microplastics became a major area of scientific focus.

As a result, the ability to capture extremely small plastic particles can vary depending on:

- treatment methods
- filtration capability
- infrastructure condition
- particle size

Infrastructure Systems Influence Outcomes

In manufacturing, final product quality depends on every stage of the process, not one isolated step.

The same is true for municipal water systems.

Source quality, treatment methods, maintenance, and infrastructure condition all influence what ultimately reaches the consumer.

Water treatment systems are designed to improve safety and remove contaminants.

However, many systems were not originally designed specifically to remove extremely small plastic particles.

Some particles may be captured during treatment. Others may remain.

This is one reason why testing results can vary from one location to another.

Tap water travels through a complex system of sources, treatment facilities, pipes, and infrastructure before reaching the consumer.

Comparing the Systems

There is no perfect system.

Bottled water reduces reliance on aging infrastructure and provides portability and consistency.

Tap water reduces dependence on single-use packaging and long-term storage in plastic containers.

Each system has tradeoffs.

Bottled Water

- direct contact with packaging
- influenced by storage and handling
- portable and controlled

Tap Water

- influenced by source water and infrastructure
- less packaging contact
- dependent on treatment systems and plumbing

Understanding those differences is more useful than trying to declare one universally better.

Household Factors

Consumers also influence the system after purchase or collection.

For bottled water:

- storage conditions matter
- heat exposure matters
- repeated reuse of bottles changes wear patterns

For tap water:

- household plumbing matters
- home filtration systems matter
- storage containers matter

This means exposure is influenced not only by production systems, but also by consumer behavior.

Everyday Habits Influence Exposure

Two families may purchase the same bottled water or receive water from the same municipal source, yet their exposure conditions can differ significantly based on everyday habits.

One family stores bottled water in a cool pantry and uses bottles only once. Another leaves cases of water in a hot garage or vehicle for extended periods and repeatedly refills the same bottles over time. Heat, sunlight, and physical wear can all influence how packaging materials age during use.

The same principle applies to tap water.

One home may have newer plumbing and a well-maintained filtration system, while another may have older pipes, worn fixtures, or no filtration at all. Even the type of container used to store water in the refrigerator can influence the overall system.

This highlights an important point: microplastic exposure is not controlled by a single factor.

It is influenced by the combined effect of:

- product design
- infrastructure systems
- storage conditions
- consumer handling and use patterns

In many cases, the way products are stored, handled, and reused becomes part of the exposure system itself.

Small Behaviors Become System Effects

A single bottle stored in a warm environment may seem insignificant.

Across millions of consumers and repeated habits, small behaviors become part of a larger system effect.

Filtration Systems

Many consumers use filtration systems to improve water quality.

These systems vary widely in design and performance.

Some are intended primarily for:

- taste
- odor
- chlorine reduction

Others are designed to capture smaller particles.

Not all filtration systems are equally effective for microplastics. Performance depends on:

- filter type
- pore size
- maintenance
- replacement frequency

Understanding what a filtration system is designed to do is important.

Filtration can reduce exposure in some situations, but it is not a universal solution.

A Practical View

Water is essential and unavoidable.

The goal is not perfection. It is awareness and reasonable reduction where practical.

Examples include:

- avoiding unnecessary heat exposure for bottled water
- understanding local water infrastructure
- maintaining household filtration systems properly
- using appropriate reusable containers

Small, consistent habits are more useful than extreme responses.

A System Perspective

Water reflects the systems surrounding it.

Those systems include:

- packaging
- infrastructure
- storage
- transportation
- consumer behavior

Microplastics in water result from many interacting factors rather than a single source.

Understanding those interactions helps consumers make more informed choices without unnecessary fear or confusion.

Key Takeaways

- Both bottled and tap water can contain microplastics
- Bottled water involves continuous contact with packaging materials
- Tap water reflects source, treatment, and infrastructure systems
- Heat, storage, and handling influence outcomes
- There is no perfect system, only different tradeoffs

Discussion Questions

1. How do bottled and tap water systems differ?
2. Why do storage conditions matter for bottled water?
3. How does infrastructure influence tap water quality?
4. Why is it difficult to declare one system universally better?

Application Exercise

Evaluate your primary source of drinking water.

Identify:

- how it is stored
- how it is transported or handled
- whether heat exposure occurs
- whether filtration is used

Then identify one practical change that could reduce unnecessary exposure or improve handling practices.

Chapter 4: Food and Packaging

Food and packaging are closely connected.

Most food people buy, store, or consume has been in contact with packaging at some point. That packaging is designed to protect the product, maintain quality, and allow it to move safely through production, transportation, and storage.

Plastic became widely used in food packaging because it performs well.

It is:

- lightweight
- durable
- moisture resistant
- flexible
- cost-effective

It also helps reduce spoilage and damage.

Without modern packaging systems, many foods would have shorter shelf lives, higher transportation losses, and greater risk of contamination.

These are important benefits.

At the same time, packaging creates direct and continuous contact between food and material.

That contact is where microplastics become part of the discussion.

Packaging Beyond the Container

Packaging is not simply a container.

It is part of a larger process that includes:

- filling
- sealing
- transporting
- storing
- displaying
- opening
- handling
- and disposal

At each stage, the package experiences movement, pressure, temperature changes, and repeated contact.

Most of the time, packaging performs exactly as intended.

Over time, however, materials can experience gradual wear and interaction.

This is especially true when packaging is:

- flexed repeatedly
- exposed to heat
- opened and resealed
- stored for extended periods

Packaging Is Designed to Perform

In manufacturing, packaging is engineered to survive real-world handling. It must protect the product during filling, stacking, shipping, and retail display.

A package that tears, leaks, or loses shape creates immediate operational problems.

Plastic materials are often selected because they consistently meet these performance requirements.

Long-term fragmentation into microscopic particles was historically not the primary design focus.

Food Contact Materials

Different foods require different types of packaging.

Examples include:

- rigid containers
- flexible pouches
- films and wraps
- trays with sealed lids
- bottles and caps

Each material behaves differently depending on:

- temperature
- moisture
- acidity
- storage time
- handling conditions

This means there is no single packaging behavior that applies to all products.

A flexible snack pouch behaves differently from a rigid beverage bottle. A refrigerated container behaves differently from a microwavable tray.

Conditions vary widely across products and packaging types.

Different Packaging for Different Conditions

In food manufacturing, packaging is selected based on how the product will be used and distributed.

A frozen meal tray must tolerate low temperatures and reheating. A snack wrapper must resist tearing during shipping and handling. A beverage bottle must survive transportation pressure and repeated movement.

The packaging system is designed around the expected environment the product will experience.

Consumers often see only the final package on a store shelf, but each design reflects a series of performance decisions made throughout the supply chain.

Different foods require different packaging systems designed for storage, handling, transportation, and product protection.

Heat and Food Packaging

Heat plays an important role in material interaction.

In manufacturing environments, materials are routinely tested under different temperatures because heat changes material behavior.

The same principle applies in everyday consumer use.

Examples include:

- microwaving packaged food
- pouring hot liquids into containers
- leaving packaged food in warm environments
- storing products in direct sunlight

Heat can increase movement at material surfaces and influence how packaging interacts with food.

This does not automatically mean harmful exposure occurs. It means the interaction between material and contents changes under different conditions.

Temperature Influences Material Behavior

In packaging production, temperature control is critical because plastics respond differently under heat.

Consumers experience the same principle in daily life, often without noticing it.

Materials that appear stable at room temperature may behave differently when heated repeatedly or exposed to prolonged warmth.

Processing and Transportation

Food packaging systems extend far beyond the consumer.

Before reaching a kitchen or refrigerator, food may travel through:

- processing plants
- automated packaging lines
- warehouses
- trucks
- distribution centers

- retail environments

Each stage involves:

- motion
- pressure
- vibration
- stacking
- temperature variation

Packaging is designed primarily around efficiency, safety, consistency, and product protection. Long-term fragmentation historically received far less attention.

Modern Food Systems Depend on Speed and Efficiency

Automated packaging lines move products rapidly through filling, sealing, labeling, stacking, and shipping processes.

Packages may experience vibration during truck transportation, compression during pallet stacking, and temperature changes during warehousing and delivery.

Modern packaging systems are designed primarily to maintain efficiency, product protection, and shelf life across these environments.

Food packaging systems are designed to withstand movement, stacking, transportation, and storage throughout the supply chain.

Manufacturing Focuses on Immediate Performance

In production environments, decisions are usually driven by questions such as:

- Does the package protect the product?
- Does it run efficiently on equipment?
- Does it maintain shelf life?
- Does it survive shipping?

These are practical and necessary concerns.

What happens years later after disposal or repeated handling has historically received less attention.

Everyday Consumer Behavior

Consumer behavior also becomes part of the system.

Examples include:

- reheating leftovers in containers

- reusing disposable packaging
- storing food for long periods
- repeatedly opening and resealing packages

These actions may increase wear or interaction over time.

Individually, the effects may be small.

Repeated daily across millions of households, they become part of a larger system pattern.

Everyday consumer habits such as reheating, reusing, storing, and resealing packaging become part of the overall food contact system.

Not All Exposure Is Equal

One of the most important points is this:

Not all packaging creates the same level of interaction or exposure.

Outcomes depend on:

- material type
- product design
- temperature

- frequency of handling
- storage conditions
- length of contact time

This complexity is one reason why broad generalizations can be misleading.

Systems Rarely Have One Simple Cause

In manufacturing systems, outcomes usually result from multiple interacting variables rather than one isolated factor.

The same is true for food packaging and microplastics.

Material behavior, temperature, handling, and time all interact together.

A Practical View

Packaging plays an important role in modern life.

It protects food, reduces spoilage, supports sanitation, and allows efficient transportation and storage.

The goal is not to eliminate packaging entirely.

The goal is to understand where interaction is more likely and make reasonable choices where practical.

Examples include:

- avoiding unnecessary heat exposure
- using appropriate containers for reheating
- reducing prolonged storage in damaged packaging
- handling reusable containers properly

Small habits applied consistently can reduce unnecessary interaction over time.

A System Perspective

Food packaging reflects decades of engineering and design focused on performance, efficiency, and protection.

Microplastics are part of the long-term tradeoffs associated with those systems.

Understanding the issue requires balancing both realities:

- the benefits packaging provides
- and the challenges that come with material persistence and fragmentation over time

That balance is where informed decision-making begins.

Key Takeaways

- Packaging is designed to protect food and support distribution systems
- Plastic is widely used because it performs efficiently and consistently
- Heat, handling, and storage influence material interaction
- Different packaging materials behave differently under different conditions
- Practical awareness is more useful than extreme avoidance

Discussion Questions

1. Why did plastic become widely used in food packaging systems?
2. How does heat influence packaging behavior?
3. Why is it difficult to generalize about all food packaging?
4. How do consumer habits become part of the system?

Application Exercise

Choose a packaged food product you use regularly.

Identify:

- the type of packaging used
- whether heat is involved during use
- how often the package is handled or reused
- how long the food remains in contact with the material

Then identify one practical change that could reduce unnecessary interaction without making daily life more difficult.

Chapter 5: Air, Dust, and Clothing

When most people think about microplastics, they usually think about water bottles, food packaging, or waste in the environment.

Air is often overlooked.

At the same time, airborne particles may be one of the most continuous forms of exposure in everyday life.

People breathe indoor and outdoor air constantly. That air contains particles from many different sources, including dust, fibers, soil, smoke, pollen, and synthetic materials.

Some of those particles are microplastics.

Microplastics in the Air

Microplastics are present in both outdoor and indoor air.

Outdoor sources include:

- tire wear from vehicles
- breakdown of outdoor plastics
- industrial activity
- construction and infrastructure wear

Indoor environments are different.

Most indoor particles come from the materials people live with every day:

- clothing
- carpets
- upholstered furniture
- curtains
- household products

Indoor air reflects the environment people create around themselves.

Material Wear Happens Everywhere

Walking across carpet, sitting on furniture, or handling fabrics creates friction and movement that slowly releases fibers and particles over time.

Indoor Environments

Most people spend much of their time indoors.

That makes indoor exposure especially important.

Inside homes, offices, schools, and vehicles, particles are constantly being generated and redistributed.

Particles settle as dust, then become airborne again through movement and airflow, creating a continuous indoor cycle.

This is one reason indoor dust has become an important area of study.

Dust is not simply dirt from outside. It is a mixture of particles from many indoor and outdoor sources.

Microplastics can become part of that mixture.

Indoor Spaces Constantly Redistribute Particles

In homes and offices, everyday movement continuously redistributes particles.

Opening doors, walking across floors, sitting on furniture, turning on fans, or adjusting heating and cooling systems can move settled dust back into the air.

Most of this activity is invisible during normal daily life, but sunlight through windows often reveals how many airborne particles are already present indoors.

Clothing and Synthetic Fabrics

Clothing is one of the largest everyday sources of synthetic fibers.

Many fabrics are made from or blended with synthetic materials such as:

- polyester
- nylon
- acrylic
- spandex

These materials are widely used because they are:

- durable
- lightweight
- flexible
- resistant to wrinkles and moisture

These are practical advantages.

At the same time, synthetic fabrics release small fibers during normal wear.

This happens through:

- movement
- stretching
- rubbing against surfaces
- washing and drying

The release is gradual and often invisible.

Performance and Tradeoffs

Synthetic fabrics became popular because they solved practical problems. They are durable, affordable, and easy to maintain.

Those same performance qualities also mean they persist over time and release fibers gradually during use.

The tradeoff is similar to many modern material systems: performance improves, but long-term persistence increases.

Laundry and Wastewater

Laundry plays a major role in fiber release.

During washing:

- fabrics are agitated
- fibers loosen from the material
- particles enter wastewater systems

Some fibers are captured during wastewater treatment. Others may pass through into larger water systems.

Drying clothing also creates visible lint, which is another form of material release.

Most people have seen lint in a dryer filter without thinking about it as part of a larger material system.

It is a visible example of gradual fiber separation.

Laundry Systems Create Continuous Fiber Movement

Modern laundry systems are designed to clean fabrics efficiently through water, detergent, agitation, and heat.

As fabrics move against each other during washing and drying, small fibers gradually separate from the material.

Visible dryer lint is one of the easiest examples of material wear most people encounter in everyday life.

Dryer lint is a visible example of fibers separating from fabrics during normal household use and laundry cycles.

Dust as a Carrier

Dust acts as a carrier for many types of particles.

It collects on:

- floors
- shelves
- furniture
- electronics
- air vents

When people walk, clean, vacuum, or move through a room, dust becomes airborne again.

This creates repeated opportunities for inhalation and surface contact.

Microplastics can move through indoor environments this way without being noticed.

Ventilation Systems Move More Than Air

Air handling systems are designed to circulate and filter air throughout buildings.

At the same time, airflow also moves fine particles through rooms, vents, hallways, and shared indoor spaces.

This is one reason HVAC filtration and airflow design play an important role in indoor air quality management.

Residential air handling systems continuously circulate indoor air, making filtration and airflow important parts of indoor particle management.

Particles Move Through Systems

Particles released in one area can move throughout a home or building through airflow, ventilation, and everyday activity. Like dust, they may settle on surfaces and later become airborne again through movement and circulation.

Inhalation and Exposure

People inhale particles constantly as part of normal breathing.

This does not automatically mean those particles create harm.

It does mean exposure occurs through more than food and water alone.

Researchers are studying:

- how airborne microplastics move through the body
- whether different particle sizes behave differently
- how long particles remain suspended in air
- which indoor environments create higher exposure levels

Research is still developing, but awareness of airborne exposure has increased significantly in recent years.

Researchers continue studying how airborne particles move through indoor environments and interact with normal human breathing and exposure pathways.

Outdoor vs Indoor Air

Outdoor air contains microplastics too, but outdoor environments are more open and dispersed.

Indoor environments are more enclosed.

That means:

- particles may accumulate more easily
- airflow redistributes them repeatedly
- exposure can become more concentrated over time

This does not mean indoor spaces are unsafe.

It means they operate differently than open outdoor systems.

A Practical View

It is not realistic to eliminate airborne particles from daily life.

The goal is awareness and reasonable management.

Practical considerations include:

- reducing excessive dust buildup
- improving airflow and ventilation
- maintaining HVAC filters
- washing fabrics appropriately
- understanding how materials behave over time

These are normal maintenance habits, not extreme responses.

Management Rather Than Elimination

In manufacturing, the goal is rarely absolute elimination of particles. The goal is usually control, reduction, and management within acceptable conditions.

The same approach is more realistic in everyday environments.

Reduction is practical. Perfection is not.

A System Perspective

Air, dust, and clothing are closely connected parts of everyday indoor environments.

They reflect:

- material choices
- indoor environments
- ventilation
- consumer habits
- textile design

Microplastics in the air are not caused by one product or one activity.

They result from the interaction of materials and daily life over time.

Understanding that makes the issue easier to approach in a practical and balanced way.

Key Takeaways

- Microplastics are present in both indoor and outdoor air
- Indoor environments are often more concentrated than outdoor environments
- Clothing and textiles are major sources of synthetic fibers
- Dust acts as a carrier and redistribution system for particles
- Practical management is more realistic than complete elimination

Discussion Questions

1. Why are indoor environments important when discussing microplastics?
2. How do synthetic fabrics contribute to airborne particles?
3. What role does dust play in exposure pathways?
4. Why is reduction a more realistic goal than elimination?

Application Exercise

Evaluate one indoor space where you spend significant time.

Identify:

- major sources of fabric or synthetic materials
- areas where dust collects
- airflow and ventilation patterns
- cleaning or maintenance habits

Then identify one practical step that could improve airflow or reduce dust accumulation over time.

Chapter 6: What Scientists Know So Far

Microplastics are now being studied around the world.

Research is expanding rapidly across multiple fields, including:

- environmental science
- toxicology
- medicine
- public health
- materials engineering

As scientific interest has grown, public awareness has grown with it.

News reports, documentaries, government agencies, and research institutions now discuss microplastics regularly. At the same time, many people are left trying to separate established findings from speculation.

That distinction matters.

Some things are now well established.

Other questions are still being investigated.

Understanding the difference is important for maintaining a balanced perspective.

What Is Clearly Established

One fact is no longer in serious dispute:

Microplastics are widespread.

They have been detected in:

- oceans and rivers
- drinking water
- soil

- air and household dust
- food products
- wildlife
- and human samples

Research from multiple countries has confirmed their presence across environmental and biological systems.

This does not automatically define the level of health risk.

It does establish that exposure exists.

Improved Detection Changes What Scientists Can See

Advances in scientific measurement are allowing researchers to detect increasingly smaller particles in air, water, food, dust, and biological samples. Many particles that were difficult or impossible to measure years ago can now be identified more effectively.

This does not necessarily mean microplastics suddenly appeared. In many cases, improved tools are making existing exposure more visible.

Researchers now understand that exposure exists across many environments. At the same time, important questions are still being studied, including:

- which particle sizes matter most
- how particles behave in the body
- how long exposure may last
- and what levels may create meaningful health effects

Improved detection increases awareness, but research into long-term impact is still developing.

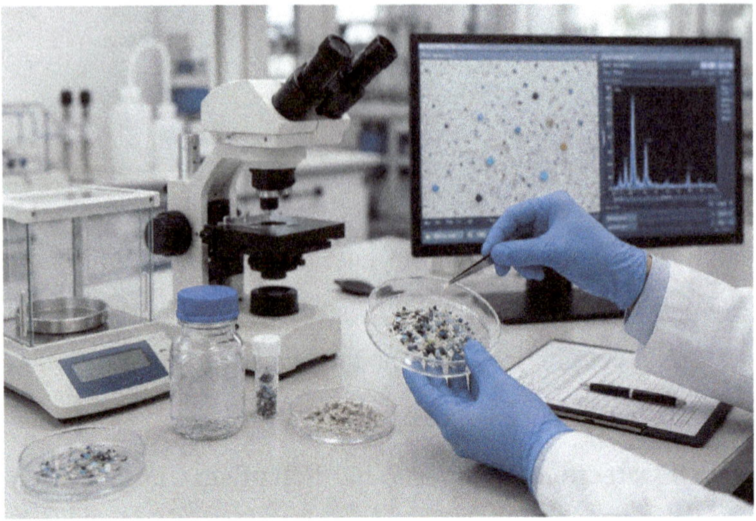

Scientists use increasingly advanced testing and measurement systems to identify and analyze microscopic particles in environmental and biological samples.

Human Exposure Is Real

Scientists now understand that people are exposed to microplastics through multiple pathways.

These include:

- drinking water
- food
- airborne particles
- indoor dust
- everyday material contact

This exposure is ongoing because microplastics are present in several interconnected systems at the same time.

Researchers have also identified microplastics in human samples, including:

- blood

- lung tissue
- placental tissue

These findings have increased scientific and public interest.

At the same time, it is important to separate detection from proven health impact.

Finding a particle in the body does not automatically mean it causes harm.

This is one of the most misunderstood parts of the discussion.

Microplastics and Endocrine Disruptor Research

One area receiving increasing scientific attention involves endocrine disruptors.

The endocrine system helps regulate hormones involved in:

- growth
- metabolism
- reproduction
- development
- and many other biological functions

Researchers are studying whether some chemicals associated with plastics may interact with these systems under certain conditions.

This topic is often misunderstood.

The concern is not simply that a visible plastic particle exists.

In many cases, the focus involves:

- chemicals used in some plastics
- additives used during manufacturing
- and how long-term exposure may occur through food, water, air, dust, and consumer products

Some chemicals used in plastics and consumer products have been identified as known or suspected endocrine-disrupting chemicals.

Examples often discussed in scientific research include:

- certain phthalates
- bisphenols such as BPA
- some flame retardants
- PFAS and related compounds

Researchers are also studying whether microplastics themselves may act as carriers for other environmental contaminants.

This is an active and evolving area of research.

Scientists continue to study:

- exposure levels
- long-term effects
- cumulative exposure
- low-dose interactions
- and how multiple chemicals may interact together over time

One challenge is that people are rarely exposed to a single substance in isolation.

Modern environments contain continuous low-level exposure from many sources, including:

- packaging
- textiles
- dust
- water systems
- personal care products
- household materials

- and industrial systems

This makes scientific evaluation more complex.

Another important distinction is that the presence of microplastics does not automatically mean harm will occur.

Research in this field is still developing.

Some studies suggest possible links between certain chemical exposures and effects involving:

- metabolism
- reproductive systems
- hormone regulation
- inflammation
- and developmental processes

At the same time, many questions remain under active investigation.

This book does not attempt to make medical conclusions.

Instead, the goal is to help readers understand why scientists are studying these systems and why the discussion around plastics has expanded beyond visible waste and litter alone.

Increasingly, the conversation includes:

- long-term exposure
- product design
- material choices
- and how modern consumer systems interact with human health and the environment

A balanced understanding is more useful than either panic or dismissal.

Practical awareness allows consumers to make informed decisions without assuming every interaction with plastic materials represents immediate danger.

Presence Does Not Automatically Mean Harm

Many public discussions move quickly from detection to conclusions about health effects.

Science usually moves more carefully.

Researchers are still studying:

- what levels of exposure matter
- whether different particle sizes behave differently
- how long particles remain in the body
- whether some materials create greater concern than others
- how the body responds over long periods of time

These are difficult questions.

Different plastics have different chemical structures and physical properties. Small fibers may behave differently from rigid fragments. Extremely small particles may interact differently than larger particles.

This complexity is one reason why scientific conclusions take time.

Different materials can behave differently under varying environmental and biological conditions. This is one reason researchers avoid broad conclusions too quickly.

One plastic may tolerate heat well but crack under repeated stress. Another may resist impact but degrade more quickly under sunlight or chemical exposure.

Engineers evaluate materials based on:

- size
- structure

- flexibility
- temperature response
- chemical resistance
- and long-term durability

The same principle applies to microplastics research.

Scientists are studying whether different particle sizes, shapes, and material types behave differently inside environmental and biological systems.

This is one reason researchers avoid making broad conclusions too quickly. Different materials and exposure conditions may produce different outcomes.

Nanoplastics and Emerging Research

Scientists are increasingly studying nanoplastics, which are even smaller than microplastics.

Because of their size, researchers are investigating whether they:

- move differently through the environment
- enter cells more easily
- behave differently inside the body

Research in this area is still developing.

Detection itself remains difficult because nanoplastics are extremely small.

This is one reason scientific uncertainty remains significant in some areas.

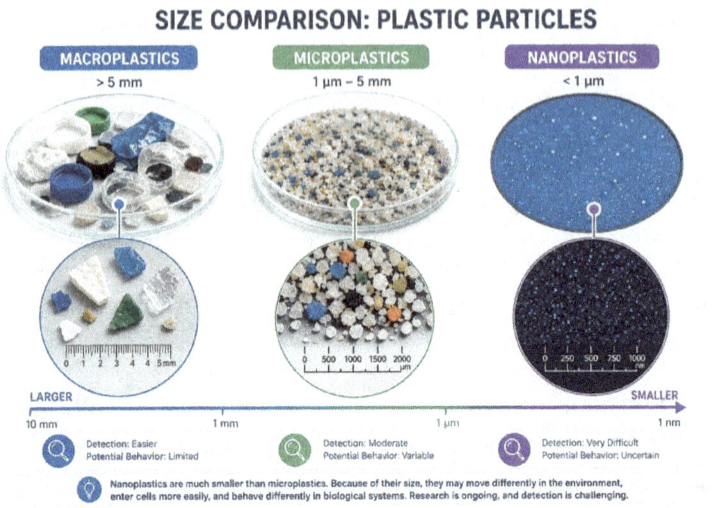

Microplastics and nanoplastics differ greatly in size, detection difficulty, and potential behavior within environmental and biological systems.

Challenges in Scientific Research

Studying microplastics is difficult for several reasons.

One challenge is consistency.

Different studies may use:

- different sampling methods
- different detection thresholds
- different particle definitions
- different laboratory procedures

This can make comparisons difficult.

Another challenge is scale.

Microplastics vary widely in:

- size
- shape

- composition
- environmental behavior

This makes it difficult to create simple conclusions that apply equally to all situations.

Complex Systems Rarely Produce Simple Answers

In manufacturing and engineering, systems that appear straightforward often become more complicated once multiple variables interact together.

A material may perform well in one environment but fail under different temperatures, stress levels, humidity, or processing conditions.

The same challenge exists in microplastics research.

Scientists must evaluate:

- different material types
- particle sizes and shapes
- environmental conditions
- exposure pathways
- and biological responses

Because these variables interact together, studies may produce different findings depending on how research is conducted.

This is one reason scientific conclusions often develop gradually over time rather than through one single study or headline.

Scientific Caution Matters

Scientific caution is sometimes misunderstood as uncertainty or indecision.

In reality, caution is part of how science works.

Researchers test findings repeatedly, compare results, and refine conclusions over time.

This process can feel slow compared to public discussion or media coverage.

People naturally want clear answers:

- Is it dangerous?
- How much exposure matters?
- What should people avoid?

Scientists are still working toward many of those answers.

That does not mean concerns should be dismissed.

It also does not mean every alarming claim should be accepted immediately.

Balanced understanding requires acknowledging both concern and uncertainty at the same time.

Public Awareness and Policy Interest

Governments, universities, and health organizations are paying increasing attention to microplastics.

Research funding has increased. Testing methods are improving. Public interest continues to grow.

Microplastics are also becoming part of broader discussions involving:

- packaging
- drinking water
- recycling systems
- consumer products
- environmental policy

This attention will likely continue increasing over the next decade.

As research develops, some current assumptions may become stronger. Others may change as more evidence becomes available.

That is normal in scientific progress.

A Practical View

For consumers, several things can be true at the same time:

- exposure exists
- research is ongoing
- uncertainty remains
- practical reduction strategies still make sense

This book does not attempt to predict future scientific conclusions.

Instead, it focuses on understanding:

- how exposure occurs
- where interaction is more likely
- and what practical steps can reasonably reduce unnecessary exposure

Fear is not necessary for awareness.

A balanced understanding is more useful than either panic or dismissal.

Awareness Often Comes Before Full Understanding

In manufacturing and public health, awareness of potential issues often develops gradually as research and real-world observations accumulate over time.

Companies may improve ventilation, update safety procedures, or modify material handling practices before every long-term effect is fully understood.

These decisions are usually based on practical risk management rather than panic.

The same approach applies to microplastics.

People do not need perfect scientific certainty to make reasonable decisions about reducing unnecessary exposure where practical.

Awareness, moderation, and informed decision-making can exist alongside ongoing scientific research.

A System Perspective

Scientific research helps explain what is being detected and studied.

The larger system helps explain why microplastics are now widespread.

That system includes:

- material design
- manufacturing
- packaging
- transportation
- consumer behavior
- waste management

Understanding both the science and the system creates a more complete picture.

Neither can be understood fully in isolation.

Key Takeaways

- Microplastics are now widely detected in environmental and human systems
- Human exposure through water, food, air, and dust is established

- Researchers are also studying chemicals associated with plastics and possible endocrine-related effects
- Detection does not automatically mean proven harm
- Research is still evolving, especially regarding long-term exposure and biological effects
- Scientific caution and public concern can exist at the same time

Discussion Questions

1. Why is it important to separate detection from proven harm?
2. How does improved testing influence public awareness?
3. Why are different particle sizes important in research?
4. Why are scientists studying chemicals associated with plastics in addition to the particles themselves?
5. Why does scientific research often move more slowly than public discussion?

Application Exercise

Find a recent article or news report about microplastics or endocrine disruptors.

Identify:

- what claims are clearly supported by evidence
- what remains uncertain
- whether the article distinguishes between detection and proven health impact
- whether the article presents a balanced or overly simplified view of risk

Then evaluate whether the overall tone feels:

- balanced

- overstated
- or dismissive

Chapter 7: What Scientists Still Don't Know

Microplastics are now widely detected in the environment and in human exposure pathways.

At the same time, many important scientific questions remain unanswered.

This is not unusual for an emerging area of research.

Public awareness often grows faster than scientific certainty. As new detection methods improve, researchers are able to identify particles more frequently and in more locations. Understanding what those findings mean takes more time.

That gap between detection and full understanding is where much of the current discussion exists.

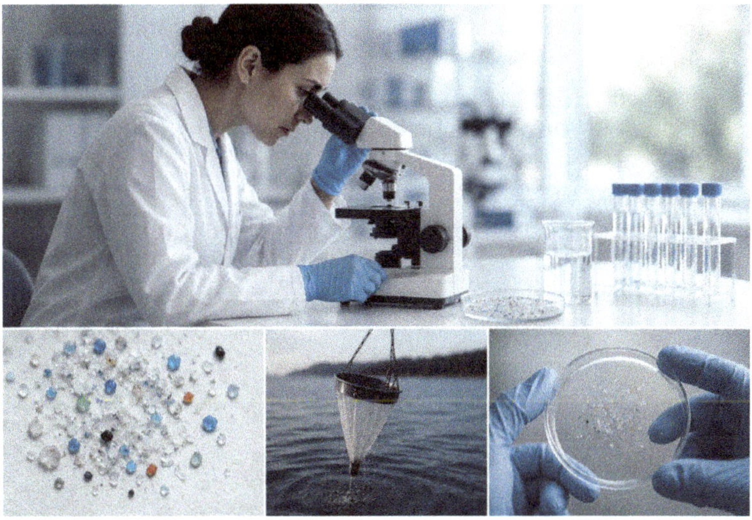

Researchers continue studying how microplastics and nanoplastics move through environmental and biological systems as detection methods and scientific understanding continue to improve.

The Difference Between Presence and Risk

One of the most important unanswered questions is this:

What level of exposure creates meaningful health risk, if any?

Scientists have confirmed that microplastics are present in:

- water
- food
- air
- dust
- and human samples

What remains less clear is:

- how different particles behave in the body
- how long they remain
- whether certain exposure levels matter more than others
- and whether some particles create greater concern than others

These are difficult questions because microplastics are not all the same.

Different materials may behave differently depending on:

- size
- shape
- chemical composition
- concentration
- exposure duration

This complexity makes broad conclusions difficult.

Long-Term Exposure

Another major question involves long-term exposure.

Microplastics are now part of modern environmental systems. This means exposure may occur gradually over many years rather than through one large event.

Researchers are studying:

- whether particles accumulate over time
- whether the body removes some particles naturally
- whether chronic low-level exposure matters
- and whether exposure during certain life stages is more important than others

Long-term studies take time.

This is one reason why scientific conclusions continue evolving.

Nanoplastics Remain Difficult to Study

Nanoplastics introduce additional uncertainty.

Because they are much smaller than microplastics, researchers are investigating whether they may:

- move differently through biological systems
- enter cells more easily
- interact differently with tissues and organs

At the same time, nanoplastics are difficult to detect consistently.

Testing technology is still improving. Different laboratories may use different methods and definitions, making comparisons challenging.

This area of research is likely to expand significantly in the coming years.

Smaller Can Be Harder to Measure

In manufacturing, the smallest defects are often the hardest to find.

A product may look acceptable to the eye, but a tiny crack, thin spot, surface flaw, or contamination issue can still affect performance. Detecting those problems requires better instruments, consistent testing methods, and experience interpreting the results.

Nanoplastics create a similar challenge.

Because they are extremely small, they are harder to detect, measure, and compare across studies. As testing methods improve, researchers may learn much more about where nanoplastics are found and how they behave.

Measurement Often Improves Before Understanding

In engineering and manufacturing, better measurement tools often reveal problems or variations that were previously invisible.

Understanding the significance of those findings usually takes additional time and testing.

The same pattern exists with nanoplastics research today.

Differences Between Studies

One reason scientific discussions can appear inconsistent is that studies are often measuring different things.

Variables may include:

- particle size
- material type
- concentration levels
- sampling methods
- laboratory techniques

As a result, findings from one study may not perfectly match another.

This does not mean the science is unreliable.

It means the field is still developing standardized methods and clearer comparisons.

Human Biology Is Complex

Another challenge is the complexity of the human body itself.

People differ in:

- age
- health conditions
- diet
- lifestyle
- environmental exposure
- occupation

These differences may influence how exposure occurs and how the body responds.

Scientists are also studying whether certain populations may experience different exposure patterns, including:

- infants and children
- industrial workers
- people living in highly urbanized environments

These questions are still under investigation.

Environmental Interactions

Researchers are also studying how microplastics interact with other substances in the environment.

Questions include:

- whether particles attract or carry chemicals
- how they behave in soil and water systems

- how sunlight, heat, and weather change material behavior over time

These interactions add another layer of complexity.

Microplastics do not exist in isolation. They move through larger environmental systems.

Materials Change Over Time

In manufacturing and product design, materials rarely behave exactly the same after years of exposure to heat, stress, sunlight, or repeated use.

The same principle applies to plastics in the environment.

Time and environmental conditions gradually change material behavior.

Environmental Systems Add Complexity

Researchers are studying how microplastics interact with:

- water systems
- soil
- sunlight
- heat
- and other environmental substances

Because microplastics move through larger environmental systems, scientists are still working to understand how environmental conditions may influence long-term behavior over time.

Why Uncertainty Should Be Expected

Scientific uncertainty is sometimes misunderstood as weakness.

In reality, uncertainty is a normal part of scientific progress.

Researchers often begin with:

- detection
- observation

- early patterns

Then move toward:

- repeated testing
- comparison of findings
- longer-term analysis

This process can take years or decades depending on the issue being studied.

Microplastics research is still relatively young compared to many other scientific fields.

Avoiding Two Extremes

Because uncertainty exists, public discussions sometimes move toward two extremes.

One extreme assumes:

- microplastics must already be causing major harm even without complete evidence

The other assumes:

- no concern is justified until every question is answered completely

Neither approach is especially useful.

A more balanced position recognizes:

- exposure is real
- research is ongoing
- uncertainty remains
- practical awareness still makes sense

This approach allows for informed decision-making without unnecessary fear or dismissal.

Balanced Decisions Often Happen Before Complete Certainty

Decisions are often made before every variable is fully understood. Researchers, consumers, industries, and policymakers regularly rely on the best information available at the time while scientific understanding continues to evolve.

Waiting for perfect certainty is rarely realistic. At the same time, reacting emotionally without sufficient evidence can also lead to poor decisions.

Microplastics research exists within this balance. Scientists continue studying long-term effects while people make practical decisions about precaution, exposure reduction, product design, and everyday behavior based on current knowledge.

Parents and consumers often make practical decisions using the best information available while scientific understanding continues to evolve.

A Practical View

Consumers do not need to wait for every scientific question to be resolved before making practical choices.

Simple actions may still make sense, including:

- reducing unnecessary heat exposure with plastics
- improving indoor air quality
- using products appropriately
- understanding where exposure is more likely

These actions are based on awareness, not panic.

The goal is reasonable reduction, not fear-driven behavior.

A System Perspective

Many unanswered questions about microplastics exist because the issue involves multiple overlapping systems:

- materials
- manufacturing
- environmental exposure
- biology
- consumer behavior

No single study or headline can fully explain all of those interactions.

Scientific understanding will continue improving over time.

For now, the most responsible approach is balanced awareness supported by practical thinking and continued research.

Key Takeaways

- Many important questions about microplastics remain unanswered
- Scientists are still studying long-term exposure and biological effects
- Nanoplastics introduce additional complexity and uncertainty

- Differences between studies often reflect evolving methods and measurements
- Balanced awareness is more useful than either panic or dismissal

Discussion Questions

1. Why is it difficult to determine long-term health effects?
2. How do nanoplastics add complexity to scientific research?
3. Why can different studies produce different findings?
4. Why is uncertainty a normal part of scientific progress?

Application Exercise

Review two different articles or studies about microplastics.

Compare:

- what each claims with confidence
- what each identifies as uncertain
- whether either source overstates or understates the current science

Then summarize where the two sources agree and where they differ.

Chapter 8: Where Exposure Is Likely Higher

Microplastics are now part of everyday environmental systems.

That means exposure can occur almost anywhere.

At the same time, exposure is not equal across all situations.

Some environments, habits, and activities are more likely to increase interaction with microplastics than others.

Understanding those situations helps move the discussion from abstract concern to practical awareness.

The goal is not to avoid modern life.

It is to recognize where exposure may be more concentrated and where reasonable adjustments may make sense.

Indoor Environments

Indoor environments are one of the most significant areas of exposure.

Most people spend a large portion of their time indoors:

- homes
- offices
- schools
- vehicles
- retail environments

These spaces contain large amounts of synthetic material, including:

- carpets
- furniture
- clothing
- electronics

- insulation
- flooring

Over time, these materials release fibers and particles through normal wear and use.

Because indoor spaces are enclosed, particles can accumulate and recirculate more easily than in outdoor environments.

Indoor Systems Naturally Accumulate Material

Indoor spaces contain many synthetic materials that experience constant everyday wear. Walking across carpet, sitting on furniture, washing clothing, opening packaging, and normal household activity can gradually release small fibers and particles into the indoor environment over time.

Because indoor environments are enclosed, particles can settle, recirculate, and accumulate more easily than in outdoor spaces. Limited airflow and repeated daily activity continuously redistribute dust and fine particles throughout homes, offices, schools, and vehicles.

Areas with Heavy Synthetic Material Use

Exposure may also be higher in environments where synthetic materials are used extensively.

Examples include:

- heavily carpeted spaces
- textile manufacturing environments
- areas with high levels of synthetic fabric use
- locations with significant plastic handling or processing

This does not automatically mean these environments are dangerous.

It means the opportunity for particle release is greater because of the amount of material present and the level of activity involved.

Heat and Repeated Plastic Contact

Heat often increases material interaction.

Situations that may create greater exposure include:

- microwaving food in plastic containers
- leaving bottled water in hot vehicles
- storing food in damaged or worn packaging
- repeated reheating in the same containers

Again, this is not about panic or isolated events.

It is about understanding where material interaction is more likely over time.

Heat Changes Material Behavior

In manufacturing and packaging operations, heat is one of the most important factors affecting material performance.

Higher temperatures can change:

- flexibility
- surface behavior
- material stability
- and interaction between materials

That is why packaging systems are tested under different storage and temperature conditions during product development.

The same principle applies in everyday consumer use.

Repeated heating, long-term storage in hot environments, and worn or damaged containers may increase the likelihood of material interaction over time.

Laundry and Textile Handling

Laundry is another area where exposure may increase.

Washing and drying synthetic fabrics release fibers through:

- agitation
- friction
- heat
- repeated mechanical movement

Households with large amounts of synthetic clothing, blankets, or upholstery may experience greater indoor fiber accumulation over time.

This is especially true in enclosed spaces with limited ventilation.

High Traffic Urban Areas

Urban environments may create additional exposure through:

- tire wear
- road dust
- industrial activity
- infrastructure wear
- dense transportation systems

Vehicle tires are a significant source of microscopic particle release. As tires wear down during driving, small particles are generated and dispersed into surrounding environments.

These particles may become part of:

- air
- stormwater runoff
- soil
- urban dust systems

This is one reason researchers increasingly view microplastics as both a consumer issue and an infrastructure issue.

Infrastructure Systems Constantly Experience Wear

In industrial and transportation systems, material wear is unavoidable.

Roads, tires, brake systems, equipment surfaces, and infrastructure components gradually wear down through normal use. Over time, small particles are continuously generated through friction, movement, weather exposure, and heavy traffic.

Most people never notice these particles because they are extremely small and widely dispersed.

Researchers are studying how modern transportation systems contribute to particle accumulation in:

- urban dust
- stormwater systems
- roadside soil
- and surrounding air environments

This is one reason microplastics are increasingly viewed as part of a broader infrastructure and environmental systems discussion.

Transportation Systems Generate Wear

All transportation systems create friction and wear over time.

Road surfaces wear. Tires wear. Brake materials wear.

These small material losses are part of normal operation and become distributed throughout the surrounding environment.

Aircraft tires experience intense friction and wear during landing, releasing microscopic particles into the surrounding environment as part of modern transportation systems.

Occupational Exposure

Some occupations may involve greater exposure because of the materials being handled.

Examples may include:

- textile production
- plastic manufacturing
- recycling operations
- industrial processing environments

Researchers are still studying how occupational exposure compares to general consumer exposure.

This is another area where long-term data is still developing.

Children and Early-Life Exposure

Children may experience different exposure patterns because of behavior and development.

Examples include:

- time spent on floors and carpets
- hand-to-mouth activity
- contact with toys and fabrics
- developing biological systems

Researchers continue studying whether certain life stages are more sensitive to exposure than others.

This does not mean parents should panic.

It does suggest that awareness and practical habits may be worthwhile.

Exposure Patterns Can Differ by Life Stage

In product safety and manufacturing, products intended for children are often evaluated differently than products intended for adults.

This is because children interact with their environments differently. Crawling, touching surfaces, placing objects in their mouths, and spending more time close to floors and fabrics can create different exposure patterns than those experienced by adults.

Researchers are studying whether similar differences may influence how children encounter microplastics in everyday environments.

Exposure Patterns Change with Behavior

In product systems, usage patterns often determine how materials wear over time.

The same principle applies to people.

Different behaviors, environments, and daily routines influence exposure patterns.

Damaged or Aging Materials

Older or damaged materials may release particles more easily than newer, intact materials.

Examples include:

- cracked containers
- heavily worn fabrics
- degraded outdoor plastics
- aging synthetic surfaces

As materials break down physically, fragmentation becomes more likely.

This is one reason product condition matters.

Exposure Is About Patterns

One important point is often overlooked:

Higher exposure usually comes from repeated patterns, not isolated moments.

A single event is rarely the entire issue.

Exposure patterns develop through:

- repeated habits
- repeated contact
- repeated environmental conditions

This is how most system-level effects operate over time.

Long-Term Systems Are Shaped by Repetition

In manufacturing and industrial operations, long-term wear rarely comes from one isolated event.

Equipment performance changes gradually through:

- repeated stress
- repeated friction
- repeated temperature cycles
- and repeated operating conditions

The same principle applies to environmental exposure.

Researchers often focus less on isolated events and more on repeated daily interaction over long periods of time.

Small Repeated Actions Create Larger Outcomes

In manufacturing systems, small repeated stresses eventually create visible wear and maintenance needs.

The same concept applies to material exposure in daily life.

Long-term patterns matter more than isolated events.

A Practical View

Understanding higher-exposure situations allows consumers to make more informed choices without extreme behavior.

Examples include:

- avoiding unnecessary heat exposure with plastics
- improving ventilation indoors
- maintaining clean living spaces
- replacing heavily damaged containers
- understanding how materials behave over time

These are practical adjustments, not major lifestyle changes.

The goal is awareness and reasonable reduction.

A System Perspective

Exposure patterns reflect how modern systems operate.

Those systems include:

- transportation
- packaging
- textiles
- indoor environments

- manufacturing
- consumer behavior

Microplastics are not concentrated in one place because modern materials are integrated throughout everyday life.

Understanding where exposure may be higher helps consumers focus on practical areas that matter most.

Key Takeaways

- Exposure levels are not equal across all environments and activities
- Indoor spaces often create more concentrated exposure conditions
- Heat, friction, and repeated handling increase material interaction
- Urban environments and transportation systems contribute particles through wear
- Exposure patterns develop through repeated daily behavior over time

Discussion Questions

1. Why are indoor environments important when considering exposure levels?
2. How does heat influence material interaction?
3. Why do repeated behaviors matter more than isolated events?
4. How do transportation systems contribute to microplastic exposure?

Application Exercise

Identify one environment where you spend significant time:

- home
- office
- vehicle
- school
- workshop

Evaluate:

- synthetic materials present
- airflow and ventilation
- heat exposure
- repeated material handling

Then identify one practical adjustment that could reduce unnecessary exposure over time.

Chapter 9: Everyday Choices That Make a Difference

Microplastics are now part of modern environmental systems.

That reality can leave people feeling uncertain about what to do next.

Some respond by trying to eliminate plastics completely. Others assume there is nothing meaningful individuals can do.

Neither extreme is especially practical.

The goal of this book has never been perfection.

It is awareness, reasonable reduction, and informed decision-making.

Small choices, repeated consistently over time, can influence exposure patterns without making daily life unnecessarily difficult.

Consistent Habits Matter More Than Perfection

Small practical habits repeated consistently over time are usually more effective than dramatic short-term changes that are difficult to maintain.

Simple actions such as reducing unnecessary heat exposure, improving indoor dust management, replacing heavily worn materials, and using products appropriately may gradually reduce unnecessary interaction over time.

The goal is not perfection. It is reasonable awareness, manageable adjustments, and long-term habits that fit everyday life.

Reducing Unnecessary Heat Exposure

One practical step involves heat.

Heat changes how materials behave. This is true in manufacturing systems and in daily consumer use.

Examples of reasonable adjustments include:

- avoiding microwaving food in containers not designed for heat
- not leaving bottled water in hot vehicles for extended periods
- replacing heavily damaged containers exposed to repeated heating
- allowing very hot food to cool slightly before sealing in plastic containers

These are simple awareness-based habits.

The goal is not fear of all packaging. It is understanding how heat influences material interaction.

Allowing extremely hot food to cool slightly before sealing in plastic containers may help reduce unnecessary material interaction over time.

Paying Attention to Product Condition

Condition matters.

Older or damaged products may release particles more easily than newer, intact materials.

Examples include:

- cracked food containers
- heavily scratched surfaces
- degraded outdoor plastics
- worn synthetic fabrics

Replacing heavily damaged items is often more practical than trying to eliminate all plastic products entirely.

This is especially true for products exposed to repeated heat or friction.

Improving Indoor Air and Dust Management

Indoor environments are important because people spend so much time inside them.

Practical steps may include:

- reducing excessive dust buildup
- improving airflow and ventilation
- maintaining HVAC systems
- washing fabrics regularly
- cleaning areas where dust accumulates

These are ordinary maintenance habits that also help reduce particle accumulation indoors.

Again, the goal is management, not elimination.

Maintenance Extends System Performance

In manufacturing environments, regular cleaning and maintenance help manage wear, buildup, and particle accumulation.

Homes and indoor spaces operate similarly.

Small maintenance habits influence how particles move and accumulate over time.

Being Thoughtful About Reuse

Reusable products can reduce waste and support sustainability goals.

At the same time, repeated reuse changes how materials wear over time.

This does not mean reusable products should be avoided.

It means consumers should:

- use products as intended
- monitor condition over time
- replace items that become heavily worn or damaged

Material condition is often more important than whether a product is reusable or disposable.

Understanding Clothing and Textiles

Synthetic fabrics are now common in modern life.

Practical considerations may include:

- washing full loads when possible
- avoiding excessive heat during drying
- recognizing that synthetic materials release fibers gradually over time

The goal is not eliminating all synthetic fabrics.

These materials provide important benefits including durability, flexibility, and affordability.

Awareness helps consumers understand how materials interact with everyday use.

Avoiding Fear-Based Decision Making

One of the most important practical choices is avoiding unnecessary fear.

Microplastics are now discussed frequently in headlines, social media, and online videos. Some information is balanced. Some is exaggerated.

Fear-based decision-making often leads to:

- confusion
- unrealistic expectations
- unnecessary stress
- extreme reactions that are difficult to maintain

A more useful approach focuses on:

- understanding systems
- recognizing realistic exposure pathways
- making practical adjustments where reasonable

Balanced awareness is more sustainable than constant alarm.

Good Decisions Usually Come from Balanced Information

Reasonable habits and informed decision-making are usually more sustainable than reacting to every alarming headline or online claim.

Focus on What Matters Most

Not every source of exposure matters equally.

Consumers often become distracted by highly specific or unlikely scenarios while overlooking more common patterns.

For most people, practical focus areas include:

- repeated heat exposure with plastics
- indoor dust and ventilation
- heavily worn materials
- long-term daily habits

These areas are more meaningful than trying to eliminate every possible source of contact.

Reasonable Reduction vs. Elimination

Modern life is built around complex material systems.

Complete elimination of microplastics from daily life is not realistic.

Reduction is more practical.

That distinction matters because unrealistic goals often create frustration and confusion.

Reasonable reduction allows people to:

- stay informed
- make practical improvements
- avoid unnecessary stress
- maintain perspective

This approach is more sustainable over time.

A Practical Consumer Mindset

Consumers do not need to become experts in chemistry or toxicology to make informed decisions.

A practical mindset includes:

- understanding how materials behave
- recognizing where exposure is more likely
- avoiding unnecessary extremes
- making manageable adjustments over time

This approach supports awareness without creating fear.

A System Perspective

Everyday choices are connected to larger systems involving:

- packaging
- transportation
- textiles
- manufacturing
- infrastructure
- consumer behavior

Individual actions alone will not solve the issue.

At the same time, consumer awareness influences markets, product design, and long-term behavior patterns.

Small decisions become meaningful when repeated across millions of people over time.

That is how systems gradually change.

Key Takeaways

- Practical awareness is more useful than extreme avoidance
- Small repeated habits matter more than isolated actions
- Heat, wear, and dust management are realistic focus areas
- Product condition influences material behavior over time
- Reduction is more practical than elimination

Discussion Questions

1. Why are small repeated habits important in reducing exposure?
2. How does heat influence material interaction?

3. Why is fear-based decision-making often unhelpful?
4. Why is reduction more realistic than elimination?

Application Exercise

Choose one daily habit involving food storage, laundry, water use, or indoor maintenance.

Evaluate:

- where repeated material interaction occurs
- whether heat or friction is involved
- whether simple adjustments are possible

Then identify one practical change that could realistically be maintained long term.

Chapter 10: Children, Families, and Common Concerns

Concerns about microplastics often become more personal when children and families are involved.

Parents naturally pay close attention to:

- food
- drinking water
- air quality
- household products
- and long-term health questions

As awareness of microplastics has increased, many families have started asking practical questions:

- Are children exposed differently than adults?
- Are some products more important than others?
- What changes are actually worthwhile?
- How much concern is reasonable?

These are understandable questions.

At the same time, clear answers are not always available because research is still developing.

That uncertainty can create anxiety if the issue is approached without balance.

Children Experience Different Exposure Patterns

Children interact with their environments differently than adults.

Examples include:

- spending more time on floors and carpets
- frequent hand-to-mouth activity

- close contact with toys, fabrics, and surfaces
- developing biological systems

These differences may influence exposure patterns.

Researchers are studying whether early-life exposure may differ from adult exposure in meaningful ways.

At the same time, it is important not to assume that every exposure automatically creates harm.

This distinction matters throughout the discussion of microplastics.

Products for Children Are Often Evaluated Differently

In manufacturing and product design, products intended for children are often evaluated differently than products intended for adults.

This is because children interact with products and environments in unique ways. Crawling, touching surfaces, mouthing objects, and spending more time close to floors and fabrics can create different patterns of contact and exposure.

Researchers are studying whether similar differences may influence how children encounter microplastics in everyday environments.

Children often spend more time in close contact with floors, carpets, toys, and indoor surfaces during everyday activity.

Toys and Everyday Products

Children are surrounded by modern materials from an early age.

Examples include:

- toys
- bottles and cups
- food packaging
- synthetic fabrics
- school supplies
- furniture and flooring

Most of these products are designed to meet safety standards and perform reliably in everyday use.

At the same time, repeated handling, friction, and wear gradually affect all materials over time.

This does not mean parents should fear every product.

It means awareness of material behavior can support more informed choices.

Indoor Environments Matter

Indoor environments are especially important for families because children often spend significant time indoors.

Dust accumulation, airflow, and fabric use can all influence particle movement within homes.

Practical household habits may help reduce unnecessary exposure, including:

- maintaining clean living spaces
- reducing excessive dust buildup
- improving ventilation
- replacing heavily damaged products

These are ordinary maintenance practices, not extreme interventions.

Indoor Spaces Require Ongoing Maintenance

In manufacturing and commercial buildings, indoor environments are regularly maintained to manage dust, airflow, and particle accumulation.

Ventilation systems, filters, cleaning schedules, and routine maintenance all help improve indoor conditions over time.

The same principle applies in homes.

Carpets, fabrics, furniture, and everyday household activity naturally create and collect dust. Regular cleaning and airflow management are simple maintenance habits that help reduce indoor particle buildup over time.

Food and Heating Practices

Food preparation and storage are common areas where families can make practical adjustments.

Examples include:

- avoiding unnecessary overheating of plastic containers
- using products according to intended use
- replacing heavily worn food containers
- being aware of repeated heat exposure over time

These are practical decisions based on material behavior, not fear.

Modern food systems depend heavily on packaging for safety, sanitation, and convenience.

The goal is reasonable awareness, not unrealistic elimination.

Everyday food preparation and storage habits influence how packaging and containers are used over time.

Bottled Water and Family Habits

Many families rely on bottled water for convenience, travel, or personal preference.

Others rely primarily on tap water and filtration systems.

Both systems involve tradeoffs.

Practical considerations may include:

- avoiding prolonged heat exposure for bottled water
- understanding local water systems
- maintaining household filtration systems properly
- using reusable containers in good condition

Again, the goal is not perfection.

It is awareness and practical reduction where reasonable.

Social Media and Public Anxiety

One challenge for families is the amount of alarming information now circulating online.

Microplastics are frequently discussed in:

- social media videos
- online articles
- podcasts
- headlines designed to attract attention

Some information is balanced and evidence-based.

Some is speculative or exaggerated.

Fear-based messaging can create unnecessary stress, especially for parents trying to make responsible decisions for their families.

This is why balanced understanding matters.

Complex Issues Are Often Oversimplified Online

In manufacturing, engineering, and product design, complex systems are rarely explained accurately through short headlines or simplified sound bites.

Important decisions usually require:

- context
- balanced information
- realistic tradeoffs
- and an understanding of how systems actually operate

The same challenge exists in public discussions about microplastics.

Online content is often designed to attract attention quickly, which can sometimes amplify fear more than understanding. Balanced information helps families make practical decisions without unnecessary anxiety.

Information Often Moves Faster Than Science

In many industries, public discussion moves faster than complete technical understanding.

Early information may contain assumptions, incomplete data, or exaggerated conclusions.

The same pattern can occur with emerging environmental and health topics.

Focus on Practical Habits

Families do not need to completely redesign daily life to respond thoughtfully to microplastics.

Practical focus areas may include:

- maintaining clean indoor spaces
- managing heat exposure with food containers
- replacing heavily damaged products

- improving ventilation
- understanding where exposure is more likely

These are manageable habits that fit within normal routines.

Small consistent behaviors are more sustainable than dramatic changes.

Avoiding Perfectionism

Perfectionism often creates frustration.

Microplastics are now part of modern life. Complete elimination from everyday life is not realistic.

Trying to avoid every possible exposure can quickly become overwhelming.

A more balanced approach recognizes:

- exposure exists
- uncertainty remains
- practical reduction still makes sense

This mindset allows families to stay informed without becoming consumed by fear.

Complex Systems Are Rarely Perfect

In manufacturing and industrial operations, systems are usually managed through continuous improvement rather than perfect elimination of every issue.

Companies work to:

- reduce risk
- improve processes
- minimize unnecessary waste
- and make practical improvements over time

The same principle applies to microplastics awareness.

Reasonable reduction strategies are often more practical and sustainable than attempting to control every possible exposure in daily life.

Reasonable Standards Work Better Than Unrealistic Ones

In manufacturing and quality systems, practical standards are usually more effective than impossible perfection goals.

The same idea applies in daily life.

Consistent reasonable habits are more sustainable than extreme efforts that cannot be maintained long term.

A Long-Term Perspective

Microplastics research will continue evolving for years.

New studies will likely:

- improve detection methods
- refine understanding of exposure
- clarify which concerns matter most
- and identify where practical improvements are most effective

Families do not need to wait for every scientific question to be answered before making reasonable decisions.

Balanced awareness is enough to begin.

A System Perspective

Children and families experience microplastics through the same larger systems discussed throughout this book:

- packaging
- textiles
- indoor environments
- transportation

- consumer products
- infrastructure

No single product or single habit explains the entire issue.

Exposure reflects the interaction of many systems over time.

Understanding those systems allows families to focus on practical actions that matter most without unnecessary fear or confusion.

Key Takeaways

- Children may experience different exposure patterns than adults
- Indoor environments and household habits play important roles
- Practical maintenance and awareness are more useful than fear-based reactions
- Heat, wear, and repeated handling influence material behavior over time
- Balanced understanding helps families make more sustainable decisions

Discussion Questions

1. Why might children experience different exposure patterns than adults?
2. How do household environments influence exposure?
3. Why is fear-based information especially challenging for families?
4. Why is balanced awareness more sustainable than perfectionism?

Application Exercise

Evaluate one area of your home environment where children or family members spend significant time.

Consider:

- flooring and fabrics
- ventilation and airflow
- food preparation and storage habits
- product condition and wear

Then identify one practical adjustment that could improve the environment without creating unnecessary stress or disruption.

Chapter 11: The System Behind the Problem

Microplastics are often discussed as if they are a single environmental issue.

In reality, they are the result of multiple systems interacting over time.

That distinction matters.

No single product, company, or consumer behavior fully explains why microplastics are now widespread. The issue developed gradually through decades of material innovation, manufacturing growth, transportation expansion, consumer demand, and modern convenience.

To understand microplastics clearly, it helps to step back and look at the larger system that created them.

Why Plastics Became Dominant

Plastic became widely used because it solved real-world problems.

Compared to many alternative materials, plastics offered:

- lower weight
- lower transportation cost
- durability
- flexibility
- moisture resistance
- lower breakage rates
- lower manufacturing cost

These advantages transformed modern production and distribution systems.

Plastic helped:

- extend food shelf life
- reduce shipping damage
- improve sanitation

- lower costs for consumers
- support mass manufacturing

These benefits are real.

Without plastics, many modern systems would operate very differently.

Materials Are Chosen for Performance

In manufacturing and packaging operations, materials are selected based on how well they perform within the larger system.

Factors often include:

- weight
- transportation efficiency
- product protection
- sanitation
- durability
- production speed
- and cost

A package that is lighter, stronger, and easier to transport can improve efficiency across the entire system.

Plastic became widely adopted because it consistently performed well across many of these categories at the same time.

Its widespread use was not accidental. It was driven by practical advantages that supported modern manufacturing, food distribution, healthcare, and global transportation systems.

Durability Created Long-Term Tradeoffs

The same durability that made plastics successful also created long-term consequences.

Materials designed to resist moisture, stress, and breakdown do not disappear easily after use.

Instead, they fragment gradually over time.

This is one reason microplastics are now widespread.

The system optimized for:

- performance
- efficiency
- durability
- convenience

Long-term particle behavior received far less attention during the early growth of modern plastics.

That was not necessarily negligence.

It reflected the priorities and knowledge of the time.

Scale Magnifies Small Effects

One important factor is scale.

A single plastic product or small amount of material wear may seem insignificant on its own. Modern production and distribution operate at enormous scale, involving:

- billions of packages
- billions of bottles
- billions of synthetic garments
- millions of vehicles generating tire wear daily

Small material losses repeated across these large systems over long periods of time eventually become broader environmental patterns.

This is how many large-scale environmental issues develop. Not through one dramatic event, but through continuous small releases repeated across billions of products, vehicles, and daily activities over time.

Consumer Convenience and Expectations

Consumer expectations also shaped the system.

Modern consumers often expect products to be:

- lightweight
- inexpensive
- portable
- durable
- resealable
- convenient

Plastic materials support those expectations effectively.

This created strong economic incentives for continued plastic use across industries.

At the same time, convenience systems often prioritize immediate performance over long-term material persistence.

Again, this was not unique to plastics. Many industrial systems developed this way.

Transportation and Global Distribution

Modern supply chains depend heavily on durable packaging materials. Products now move across:

- countries
- oceans
- warehouses
- distribution networks
- retail environments

During transportation and storage, packaging may experience:

- vibration
- stacking pressure
- repeated loading and unloading
- temperature changes

- weather exposure
- and extended shipping times

Materials that are lightweight, durable, and resistant to moisture and breakage help reduce product damage and transportation loss throughout these large distribution networks.

Plastic became widely used because it performed well under these demanding conditions. Without durable packaging, modern global commerce would experience significantly higher levels of product damage, waste, and logistical inefficiency.

Packaging systems are designed to survive large-scale transportation, warehousing, and distribution environments.

Recycling and Waste Systems

Waste and recycling systems also influence the larger picture.

Many plastics were originally designed for performance during use, not necessarily for easy recovery or recycling after disposal.

Over time, this created challenges involving:

- collection systems

- sorting infrastructure
- contamination
- economics of recycling
- material degradation after repeated use

Even effective recycling systems do not eliminate all fragmentation or environmental release.

This is one reason microplastics remain part of the broader discussion.

Infrastructure and Urban Environments

Microplastics are connected not only to consumer products, but also to large infrastructure and transportation environments.

Examples include:

- road wear
- tire particles
- synthetic turf
- industrial coatings
- construction materials
- water infrastructure

These materials operate continuously under conditions involving:

- friction
- weather exposure
- vibration
- repeated stress
- and long-term environmental aging

Over time, gradual wear can release small amounts of material into the surrounding environment. Because transportation networks, roads, industrial surfaces, and infrastructure operate continuously and at large scale, even small repeated losses may contribute meaningfully to broader environmental exposure patterns.

Many people rarely think about infrastructure as part of the microplastics discussion, even though these sources may play an important role in long-term environmental particle release.

Why the Issue Is Difficult to Solve Quickly

Because microplastics are tied to so many interconnected systems, simple solutions are difficult.

Replacing one product or banning one material does not automatically eliminate the broader issue.

The challenge involves:

- manufacturing
- infrastructure
- transportation
- consumer expectations
- economics
- material science
- waste management

This complexity is one reason change tends to happen gradually.

Systems built over decades usually evolve over decades.

Innovation and Future Design

Awareness of microplastics is already influencing:

- packaging design
- material research
- filtration systems
- infrastructure discussions
- consumer behavior

Future solutions may involve:

- alternative materials
- improved recycling systems
- better filtration technologies
- redesigned packaging systems
- reducing material used in certain applications

No single solution will likely solve the issue completely.

Improvement will probably come through many smaller changes across multiple systems.

Long-Term Improvement Will Likely Be Gradual

Major improvements rarely come from one single breakthrough. Progress usually occurs through many smaller changes involving:

- material improvements
- product redesign
- better filtration
- improved recycling systems
- transportation efficiencies
- and infrastructure upgrades

The same pattern will likely apply to microplastics. Long-term improvement will probably involve gradual adjustments across packaging, transportation, recycling, infrastructure, and consumer products rather than one universal solution.

A Practical Perspective

Understanding the larger system helps consumers avoid oversimplified thinking.

Microplastics are not simply the result of careless behavior.

They are connected to:

- how products are designed

- how materials are selected
- how goods are transported
- how modern life operates

This perspective supports more informed discussion and more realistic expectations.

A System Perspective

Microplastics are ultimately a systems issue.

They reflect decades of decisions involving:

- convenience
- durability
- cost
- efficiency
- and large-scale industrial growth

Understanding that does not solve the problem immediately.

It does create a clearer picture of why microplastics became widespread and why thoughtful long-term change will require system-level thinking rather than isolated reactions.

Key Takeaways

- Microplastics developed through interconnected modern systems over time
- Plastics became dominant because they solved important practical problems
- Durability and scale created long-term fragmentation challenges
- Transportation, infrastructure, and consumer expectations all contribute to the issue

- Long-term improvement will likely require gradual system-wide changes

Discussion Questions

1. Why did plastics become so widely used across industries?
2. How did durability become both a benefit and a challenge?
3. Why are system-level issues difficult to solve quickly?
4. How do consumer expectations influence material systems?

Application Exercise

Choose one everyday product system such as:

- bottled beverages
- food packaging
- clothing
- household shipping
- transportation

Identify:

- why plastics are used in the system
- what benefits they provide
- where fragmentation or wear may occur over time
- what realistic improvements might be possible in the future

Then summarize the tradeoffs involved in changing the system.

Chapter 12: A Practical Consumer Checklist

Microplastics are now part of modern environmental systems.

That reality can feel overwhelming at first.

Throughout this book, one theme has remained consistent:
the goal is not perfection. It is practical awareness and reasonable reduction where possible.

Consumers do not need to eliminate all plastics from daily life to respond thoughtfully to the issue.

Most meaningful changes come from:

- understanding how materials behave
- recognizing where exposure is more likely
- and making manageable adjustments over time

This chapter brings those ideas together into a practical consumer checklist.

The purpose is not to create fear or rigid rules.

It is to help consumers focus on realistic habits that fit into everyday life.

Food and Packaging

Be Aware of Heat Exposure

Heat changes how materials behave.

Practical habits may include:

- avoiding unnecessary overheating of plastic containers
- allowing very hot food to cool slightly before sealing in plastic storage containers
- using products according to intended use instructions
- replacing heavily worn or damaged food containers

The goal is awareness, not fear of all packaging.

Pay Attention to Product Condition

Material condition matters over time.

Products that are:

- cracked
- heavily scratched
- brittle
- or visibly worn

may release particles more easily than intact products.

Replacing damaged items when practical is often more useful than attempting to avoid all plastics entirely.

Reduce Unnecessary Long-Term Storage in Damaged Packaging

Food packaging is designed for performance during normal use and storage.

At the same time, prolonged use of heavily damaged or aging packaging may increase unnecessary material interaction.

Simple awareness helps support better habits.

Materials Perform Best Within Intended Use Conditions

Products are generally designed and tested for expected use conditions. Over time, repeated heat exposure, scratching, wear, and aging can gradually affect material performance. Replacing heavily worn or damaged containers is a practical maintenance habit based on normal material behavior over time.

Material Performance Changes Gradually with Wear

In manufacturing and industrial operations, equipment and materials are routinely inspected for wear, cracking, surface damage, and loss of performance over time.

Most products are designed to function reliably within normal use conditions. However, repeated stress, heat exposure, bending, scratching, washing, and aging can gradually affect how materials perform.

Consumers experience the same principle in daily life through repeated handling, heating, washing, and reuse of products over time.

The same principle applies to household products and food containers.

Replacing heavily worn or damaged items is a practical maintenance decision based on normal material behavior rather than fear.

Material condition changes gradually through repeated use, heat exposure, washing, and everyday handling over time.

Drinking Water

Avoid Unnecessary Heat Exposure

For bottled water:

- avoid prolonged storage in hot vehicles
- reduce unnecessary exposure to direct sunlight and heat
- store products under normal indoor conditions when possible

Heat and time can influence material interaction.

Maintain Household Filtration Systems Properly

Filtration systems only perform effectively when maintained correctly.

Practical habits include:

- replacing filters on schedule
- understanding what the system is designed to remove
- following manufacturer recommendations

No filtration system is perfect, but maintenance matters.

Understand Your Local Water System

Consumers often know little about:

- local water sources
- treatment systems
- infrastructure conditions

Basic awareness helps place information into context.

Indoor Air and Dust

Reduce Excessive Dust Accumulation

Dust acts as a carrier for many particles, including microplastics.

Practical habits include:

- regular cleaning
- maintaining ventilation systems
- improving airflow where practical
- cleaning areas where dust accumulates heavily

These are normal maintenance practices that also support healthier indoor environments generally.

Pay Attention to Ventilation

Indoor air behaves differently than outdoor air.

Improving airflow and ventilation can help reduce particle accumulation in enclosed environments.

This does not require major renovation or expensive systems in most cases.

Small improvements often help.

Airflow Management Is Important in Enclosed Systems

In manufacturing and industrial facilities, airflow and ventilation are important because enclosed environments naturally allow dust and fine particles to accumulate over time.

Ventilation systems help:

- improve air circulation
- reduce stagnant air
- manage airborne particles
- and maintain more stable indoor conditions

The same principle applies to homes, offices, schools, and vehicles.

Simple habits such as improving airflow, using ventilation fans, opening windows when practical, and maintaining HVAC systems can help reduce indoor particle accumulation over time.

Maintenance Influences Indoor Conditions

In manufacturing and building systems, routine maintenance strongly affects long-term performance.

Homes and indoor environments work similarly.

Small consistent maintenance habits gradually influence indoor particle conditions over time.

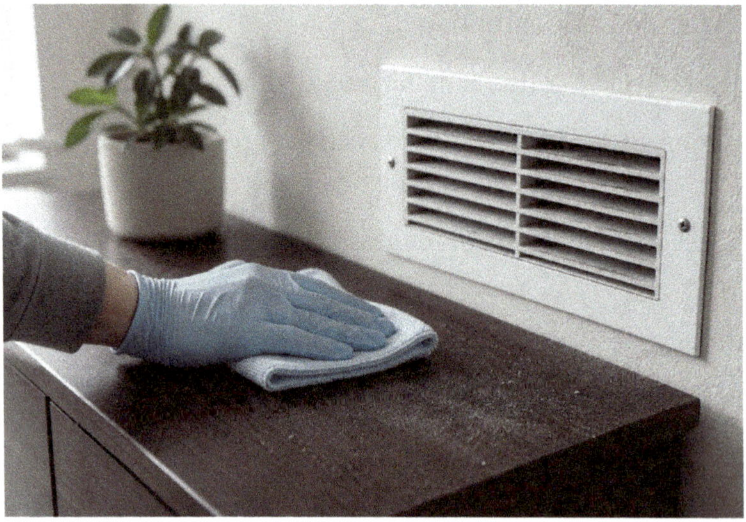

Indoor maintenance and airflow influence how dust and particles accumulate and circulate over time.

Clothing and Laundry

Recognize How Synthetic Fabrics Behave

Synthetic fabrics release fibers gradually during:

- wear
- washing
- drying
- repeated use

Understanding this behavior helps consumers make more informed decisions without needing to eliminate all synthetic materials.

Avoid Unnecessary Over Washing

Frequent washing and high heat drying increase fabric wear over time.

Reasonable laundry habits may help reduce unnecessary fiber release while also extending product life.

Pay Attention to Product Wear

Older heavily worn fabrics may release fibers more easily than newer materials in good condition.

Again, awareness matters more than perfection.

Product Use and Reuse

Use Products as Intended

Products generally perform best when used within their intended design conditions.

Improper heating, excessive wear, or repeated use beyond intended conditions may increase material degradation.

Replace Heavily Damaged Items

Not every product needs to be discarded at the first sign of wear.

At the same time, heavily degraded products may no longer perform as intended.

Practical judgment matters.

Performance Depends on Proper Use

In manufacturing systems, products operate most effectively when used within their intended performance range.

Consumers experience the same principle with everyday products and materials.

Avoid Fear-Based Decision Making

Microplastics are now a frequent topic in:

- news headlines
- social media
- online videos
- health discussions

Some information is balanced and evidence-based.

Some is exaggerated or speculative.

Consumers benefit from:

- evaluating sources carefully
- recognizing uncertainty where it exists
- separating detection from proven harm

Balanced awareness is more useful than constant alarm.

Focus on Long-Term Habits

Long-term habits matter more than isolated actions.

Practical consistent behaviors are more sustainable than extreme short-term reactions.

Examples include:

- maintaining indoor environments
- reducing unnecessary heat exposure
- replacing damaged products
- understanding material behavior over time

Small improvements repeated consistently often matter most.

Long-Term Results Usually Come From Consistency

In manufacturing and industrial operations, long-term performance is usually improved through consistent daily practices rather than one-time dramatic changes.

Systems improve through:

- routine maintenance
- repeatable procedures
- gradual process improvements
- and consistent operational habits over time

The same principle applies to reducing unnecessary exposure.

Small practical habits maintained consistently are often more effective and sustainable than extreme efforts that are difficult to continue long term.

Systems Improve Through Consistency

In manufacturing and engineering systems, steady improvement usually comes from repeated consistent actions rather than dramatic one-time changes.

The same principle applies to everyday consumer behavior.

A Balanced Perspective

Consumers do not need to become experts in chemistry or toxicology to make practical decisions.

Reasonable awareness is enough.

Several things can be true at the same time:

- microplastics are widespread
- exposure occurs regularly
- important scientific questions remain unanswered
- practical reduction strategies still make sense

Balanced understanding allows consumers to stay informed without becoming overwhelmed.

A System Perspective

Everyday choices exist within much larger systems involving:

- manufacturing
- packaging
- textiles
- transportation
- infrastructure
- consumer behavior

Individual actions alone will not eliminate microplastics.

At the same time, consumer awareness influences markets, product design, and long-term system change.

Systems often evolve gradually through millions of small decisions repeated over time.

Key Takeaways

- Practical awareness is more useful than fear-based reactions
- Heat, wear, and repeated handling influence material behavior
- Indoor maintenance and airflow matter
- Product condition influences long-term performance
- Consistent habits are more effective than extreme short-term actions

Discussion Questions

1. Why is practical awareness more sustainable than perfectionism?

2. How do small repeated habits influence long-term exposure patterns?
3. Why does product condition matter?
4. How can consumers evaluate information more critically?

Application Exercise

Review your own daily routines involving:

- food storage
- drinking water
- laundry
- indoor maintenance
- reusable products

Identify:

- one habit you already manage well
- one area where improvement is possible
- one concern that may not require significant worry after better understanding the system

Then create a simple realistic plan for one practical improvement you could maintain consistently over time.

Chapter 13: Government Response and Emerging Research

Microplastics are no longer viewed only as an environmental discussion.

Over the past several years, the topic has increasingly moved into:

- government research programs
- public health discussions
- environmental policy conversations
- drinking water studies
- and infrastructure planning

This shift matters because large institutions typically do not commit significant resources to an issue unless they believe it may become increasingly important over time.

That does not automatically mean the worst conclusions have already been reached.

It does mean awareness is expanding beyond academic research alone.

Microplastics are now being discussed at national and international levels because they intersect with:

- public health
- packaging systems
- environmental management
- water infrastructure
- manufacturing
- and long-term sustainability planning

The issue is increasingly being viewed as part of broader environmental, infrastructure, and public health systems.

Government Attention Often Signals System-Level Concern

In manufacturing and engineering, management attention usually increases when a problem appears across multiple systems at the same time.

A small isolated issue may remain localized.

A system-wide issue receives broader investigation because it may affect:

- operations
- cost
- safety
- infrastructure
- or long-term performance

The same principle applies to environmental and public health systems.

As microplastics began appearing in water studies, air sampling, biological samples, food systems, and environmental monitoring, government agencies and research institutions began expanding their attention accordingly.

This does not mean every question has been answered.

It means enough uncertainty and visibility now exist to justify larger-scale investigation.

Research Funding and Scientific Expansion

One of the clearest signs of growing concern is the increase in research funding.

Universities, health organizations, and government agencies are investing in:

- improved testing methods
- exposure studies
- toxicology research

- environmental monitoring
- filtration technologies
- and nanoplastics research

Much of this work focuses on improving scientific consistency.

Researchers are still trying to answer basic but important questions:

- How much exposure occurs?
- Which particle sizes matter most?
- How do particles behave in the body?
- What exposure levels are meaningful?
- How should microplastics be measured consistently?

These questions are difficult because microplastics vary widely in:

- size
- shape
- composition
- environmental behavior
- and exposure pathways

This complexity is one reason scientific conclusions continue evolving.

Measurement Improves Before Full Understanding

In industrial and engineering systems, organizations often improve testing and monitoring before every long-term effect is fully understood.

Better measurement helps identify:

- where potential problems exist
- how systems behave under real-world conditions
- and where additional research or investigation is needed

The same pattern is occurring with microplastics.

Improved detection expands understanding of where microplastics exist and how they move through environmental and biological systems.

The STOMP Initiative

One example of expanding federal attention is the initiative known as STOMP, launched by the Advanced Research Projects Agency for Health (ARPA-H) in April 2026.

STOMP stands for:
Systematic Targeting of Microplastics.

The initiative was launched to support expanded research into microplastics and nanoplastics and reflects growing institutional interest in understanding how these particles interact with environmental and biological systems.

The program focuses on areas including:

- improved detection methods
- exposure measurement
- toxicology research
- biological interaction studies
- and, in a later phase, possible methods for removing microplastics and nanoplastics.

Programs like STOMP represent an important transition point.

Microplastics are moving from a topic discussed primarily in environmental circles toward a broader public health and infrastructure discussion.

That shift does not mean final conclusions have been reached.

It does suggest that institutions expect the issue to remain important for many years.

Government and scientific institutions are increasing research into microplastics, nanoplastics, exposure pathways, and detection technologies.

Drinking Water and Infrastructure

Government attention has also expanded into drinking water systems.

Researchers and agencies are increasingly studying:

- source water contamination
- treatment effectiveness
- infrastructure wear
- filtration systems
- and long-term environmental accumulation

This is partly because water systems connect directly to daily human exposure.

Municipal systems were originally designed around concerns such as:

- bacteria
- pathogens
- sediment

- chemical contamination
- and traditional pollutants

Microplastics introduce a newer category of concern involving extremely small particles moving through large infrastructure systems.

Different treatment systems may capture particles differently depending on:

- filtration capability
- infrastructure condition
- particle size
- and operational design

This is one reason testing results may vary between locations.

Infrastructure Evolves Slowly

Large infrastructure systems are not redesigned overnight.

Water systems, transportation systems, and waste systems often evolve gradually over decades because:

- they are expensive
- interconnected
- heavily regulated
- and critical to everyday life

The same pattern is likely with microplastics management.

Future improvements will probably occur incrementally through:

- improved filtration
- better material design
- updated infrastructure
- and more advanced monitoring systems over time

International Attention

Microplastics are also receiving increasing international attention.

Governments and organizations around the world are studying:

- packaging systems
- synthetic textiles
- tire wear
- recycling infrastructure
- and environmental accumulation pathways

This attention overlaps with broader discussions involving:

- sustainability
- circular economy systems
- packaging regulation
- and extended producer responsibility (EPR)

As awareness grows, material systems may increasingly be evaluated not only on:

- cost
- performance
- and durability

but also on:

- long-term environmental behavior
- recyclability
- and lifecycle impact

This represents a significant shift in how modern material systems may eventually be evaluated.

System Priorities Change Over Time

In manufacturing history, systems are often optimized first for:

- speed
- cost
- efficiency
- and production performance

As understanding evolves, priorities expand.

Safety, emissions, waste management, energy efficiency, and long-term environmental performance often become more important later.

Microplastics may represent part of a similar transition.

Public Awareness and Media Attention

Public awareness has increased rapidly alongside scientific research.

Microplastics are now discussed regularly through:

- news reports
- documentaries
- social media
- podcasts
- and public policy discussions

Some coverage is balanced and informative.

Some is highly emotional or speculative.

This creates challenges for consumers trying to understand:

- what is clearly established
- what remains uncertain
- and what practical actions actually matter

Fear-based communication often spreads faster than careful scientific explanation.

That is one reason balanced education remains important.

Scientific Research Takes Time

Public attention often moves faster than scientific certainty.

Early findings may raise important questions long before researchers fully understand long-term patterns or health implications.

Researchers must still:

- test findings repeatedly
- compare methods
- validate results

- and study long-term patterns over time

This process can take years.

At the same time, public interest usually demands immediate answers.

That tension often creates confusion between:

- early findings
- preliminary concerns
- and fully established conclusions

Understanding that difference helps maintain perspective.

A Practical View

For consumers, several things can be true at the same time:

- microplastics are widespread
- government and scientific attention is increasing
- important questions remain unanswered
- and practical awareness still makes sense

People do not need to panic in order to stay informed.

Balanced awareness is more useful than either:

- fear-driven reactions

or

- complete dismissal of the issue

Practical understanding allows consumers to:

- recognize realistic exposure pathways
- evaluate information more carefully
- and make reasonable long-term decisions without unnecessary stress

Looking Ahead

Government attention to microplastics will likely continue growing over the next decade.

Future efforts may include:

- improved detection standards
- expanded research programs
- updated drinking water guidance
- infrastructure improvements
- packaging redesign
- and broader environmental monitoring

Some current assumptions may become stronger as evidence develops.

Other assumptions may change.

That is normal in scientific progress.

The important point is this:

Microplastics are no longer viewed only as a narrow environmental issue.

They are increasingly being viewed as part of a larger systems discussion involving:

- materials
- infrastructure
- public health
- manufacturing
- transportation
- and long-term environmental management

That shift is already underway.

Key Takeaways

- Government and scientific attention toward microplastics is increasing rapidly

- Research funding is expanding across health, environmental, and infrastructure fields
- STOMP reflects growing federal interest in microplastics and nanoplastics research
- Scientific understanding is still evolving despite increasing public attention
- Long-term responses will likely involve gradual system-wide changes over time

Discussion Questions

1. Why does increased government attention matter in emerging scientific issues?
2. How does improved detection influence public awareness?
3. Why do large infrastructure systems evolve slowly?
4. Why is balanced communication important during ongoing scientific research?

Application Exercise

Review a recent government, university, or scientific organization discussion involving microplastics.

Identify:

- what concerns are clearly stated
- what uncertainties remain
- what actions or research priorities are being proposed
- and whether the tone feels balanced or overly emotional

Then summarize how institutional attention toward microplastics appears to be evolving over time.

Chapter 14: Looking Ahead

Microplastics are now part of modern life.

They move through:

- water systems
- food systems
- transportation networks
- packaging
- textiles
- indoor environments
- and everyday consumer products

This reality did not develop overnight.

It emerged gradually through decades of industrial growth, material innovation, global distribution, and changing consumer expectations.

Understanding that larger context is important when thinking about the future.

Microplastics are not simply a waste issue. They are a systems issue.

That means future change will likely come through gradual adjustments across many different systems rather than one single solution or breakthrough.

The Future Will Likely Be Incremental

Large systems rarely change all at once.

Infrastructure, manufacturing, packaging, transportation, and consumer behavior evolve gradually because they are:

- interconnected
- expensive
- operationally complex
- and deeply integrated into everyday life

The same pattern is likely with microplastics.

Progress will probably occur through many smaller improvements involving:

- material design
- packaging systems
- filtration technology
- recycling infrastructure
- manufacturing methods
- and consumer awareness

Some changes may seem small individually. Over time, however, incremental improvements can significantly reshape larger systems.

Long-Term Improvement Usually Happens Gradually

In manufacturing and engineering, major progress often occurs through continuous improvement rather than dramatic overnight transformation. Systems evolve step by step over time. The same pattern is likely with microplastics and environmental systems.

Material Design Will Continue Evolving

Future packaging and product systems may increasingly consider:

- long-term material behavior
- recyclability
- environmental persistence
- and end-of-life system performance

Historically, many materials were designed primarily for:

- durability
- cost
- efficiency
- and transportation performance

Those priorities will likely remain important.

At the same time, growing environmental awareness may push industries to balance performance with broader lifecycle considerations.

This could influence:

- packaging simplification
- mono-material systems
- alternative materials
- reusable systems
- and improved recovery infrastructure

No material system is perfect. All systems involve tradeoffs.

The challenge moving forward will be balancing:

- performance
- cost
- convenience
- sustainability
- and long-term system impact

Innovation Often Follows Pressure

In manufacturing history, industries usually innovate when pressure increases from:

- cost
- regulation
- customer expectations
- supply chain disruption
- or changing market conditions

Microplastics, sustainability concerns, and environmental regulation are beginning to create that type of pressure across multiple industries simultaneously.

That pressure will likely accelerate innovation over time.

Innovation often accelerates when industries face increasing pressure from cost, regulation, sustainability concerns, and changing market expectations.

Infrastructure Will Become More Important

Infrastructure systems will likely play a growing role in future solutions.

This includes:

- water treatment systems
- filtration technologies
- recycling systems
- wastewater management
- and urban environmental controls

Many current systems were not originally designed around microplastics as a primary concern.

As understanding improves, infrastructure may gradually adapt through:

- upgraded filtration
- improved monitoring
- and new engineering approaches

These changes will likely take years or decades rather than months.

Infrastructure systems evolve slowly because they operate continuously and support essential daily life.

Consumer Awareness Will Continue Growing

Public awareness of microplastics has increased rapidly.

Consumers are increasingly asking questions about:

- packaging
- water systems
- textiles
- food contact materials
- and environmental exposure

This awareness may influence:

- purchasing decisions
- product design
- corporate sustainability strategies
- and long-term policy discussions

At the same time, awareness alone does not guarantee perfect solutions.

Consumers often want:

- affordable products
- convenience
- durability
- portability
- and performance

Those expectations continue shaping the market.

Future systems will likely involve balancing these competing priorities rather than eliminating them entirely.

Avoiding Oversimplified Thinking

One important lesson moving forward is that oversimplified solutions rarely solve complex system problems.

Microplastics involve:

- manufacturing
- transportation
- infrastructure
- consumer behavior
- waste systems
- and material science

Because so many systems overlap, simple answers are unlikely.

For example:

- replacing one material may create different environmental tradeoffs
- banning one product category may shift impacts elsewhere
- increasing durability may improve one issue while worsening another

This complexity does not mean progress is impossible.

It means thoughtful systems-level thinking is necessary.

Every Solution Creates New Tradeoffs

In engineering and manufacturing, solving one problem often introduces another challenge somewhere else in the system.

Reducing weight may decrease transportation cost but reduce durability.
Increasing durability may improve performance but complicate recycling.

The same balancing process exists in environmental systems.

Research Will Continue Evolving

Scientific understanding of microplastics is still developing.

Future research will likely improve understanding involving:

- long-term exposure
- nanoplastics
- biological interaction
- environmental accumulation
- and infrastructure performance

Some current concerns may become more significant as evidence develops.

Other concerns may eventually prove less important than originally feared.

That is how science progresses.

Improved research does not always confirm early assumptions exactly as expected.
Sometimes it strengthens them. Sometimes it changes them.

Balanced understanding allows room for both possibilities.

The Importance of Perspective

Microplastics are a serious topic. They are also part of a much larger story involving how modern industrial societies developed.

Plastic became widespread because it solved real problems:

- sanitation
- transportation efficiency
- food preservation
- affordability
- and manufacturing consistency

Those benefits remain important.

At the same time, long-term environmental persistence created new challenges that earlier systems did not fully anticipate.

Understanding both sides of that history matters.

A balanced perspective recognizes:

- the benefits modern material systems created
- and the long-term tradeoffs now becoming more visible

This perspective encourages thoughtful long-term improvement rather than fear-driven reaction.

A Practical Long-Term Mindset

Consumers do not need to solve the entire issue individually.

Practical awareness remains the most useful approach.

That includes:

- understanding how materials behave
- recognizing where exposure is more likely
- maintaining perspective
- and making reasonable adjustments where practical

Small informed choices repeated consistently across millions of people can gradually influence larger systems over time.

That is often how meaningful long-term change occurs.

Systems Change Through Repeated Decisions

In large industrial systems, change usually occurs through thousands of repeated decisions made over time:

- design changes
- operational improvements
- infrastructure upgrades
- consumer preferences
- and evolving standards

Environmental systems evolve similarly.

Small repeated improvements often become significant over the long term.

A Final Perspective

Microplastics will likely remain part of scientific, environmental, and public discussion for many years.

Research will continue evolving. Policies will continue evolving. Materials and systems will continue evolving as well.

The goal of this book has not been to predict every future outcome.

It has been to encourage:

- practical understanding
- balanced awareness
- systems thinking
- and thoughtful long-term perspective

Fear alone does not solve complex problems.

Understanding systems is a better place to begin.

Key Takeaways

- Microplastics are part of larger interconnected material and infrastructure systems
- Long-term improvement will likely occur gradually across many industries and systems
- Future innovation will involve balancing performance, cost, and environmental considerations
- Scientific understanding will continue evolving over time
- Balanced systems-level thinking is more useful than oversimplified solutions or fear-driven reactions

Discussion Questions

1. Why are large systems difficult to change quickly?
2. How do tradeoffs complicate environmental decision-making?

3. Why is systems thinking important when discussing microplastics?
4. How can consumer awareness influence long-term system change?

Application Exercise

Choose one system discussed throughout this book:

- packaging
- transportation
- water infrastructure
- textiles
- consumer products

Evaluate:

- what benefits the system currently provides
- what long-term challenges it creates
- what realistic improvements may occur in the future
- and what tradeoffs those improvements might involve

Then summarize how your perspective on microplastics has changed after viewing the issue as part of a larger system rather than a single isolated problem.

For Instructors and Courses

This book was written not only for general readers, but also for:

- educators
- students
- training programs
- discussion groups
- and professional learning environments

Microplastics are often discussed primarily as an environmental topic. In reality, the issue overlaps with many different fields, including:

- environmental science
- public health
- packaging
- manufacturing
- materials engineering
- sustainability
- consumer behavior
- infrastructure
- and public policy

Because of this, the topic works well in both academic and professional discussion settings.

A Systems-Based Approach

One of the goals of this book is to encourage systems thinking.

Rather than focusing only on headlines or isolated examples, the chapters examine how multiple systems interact over time, including:

- packaging systems
- water systems
- transportation systems
- textile systems

- consumer behavior
- and infrastructure systems

This broader perspective helps students and readers understand why microplastics are difficult to reduce through simple one-dimensional solutions.

The issue involves tradeoffs, operational realities, and long-term system design choices.

Suggested Educational Uses

This book may be useful in courses or discussions involving:

- environmental science
- sustainability
- packaging and materials systems
- public health
- engineering technology
- manufacturing operations
- supply chain and logistics
- environmental policy
- consumer behavior
- and systems thinking

The material is intentionally written in plain language to encourage discussion across varied educational backgrounds rather than only technical scientific audiences.

Discussion-Based Learning

Each chapter includes:

- key takeaways
- discussion questions
- and application exercises

These sections were designed to encourage:

- practical thinking
- systems analysis

- critical evaluation of information
- and balanced discussion

The goal is not simply memorization. The goal is helping students understand how complex real-world systems involve competing priorities and tradeoffs.

Real-World Perspective

This book approaches microplastics from both a scientific and operational perspective.

Many environmental discussions focus primarily on outcomes while spending less time discussing:

- manufacturing realities
- transportation requirements
- packaging performance
- infrastructure limitations
- and consumer expectations

Understanding those operational realities helps students evaluate why material systems developed the way they did and why future improvements may require gradual system-wide change.

Complex Systems Rarely Have Simple Answers

In manufacturing, engineering, and environmental systems, problems are rarely caused by one single factor alone.

Performance, cost, durability, infrastructure, and human behavior often interact together at the same time.

The same principle applies to microplastics.

This is one reason balanced systems thinking is important in both education and public discussion.

Encouraging Critical Thinking

Microplastics are frequently discussed through:

- headlines
- social media
- documentaries
- and rapidly changing scientific reports

Students benefit from learning how to evaluate:

- evidence quality
- scientific uncertainty
- systems tradeoffs
- and emotional versus evidence-based communication

This topic creates valuable opportunities to practice:

- critical thinking
- source evaluation
- discussion-based learning
- and long-term systems analysis

Suggested Classroom Activities

Possible classroom or discussion activities may include:

- evaluating packaging systems and tradeoffs
- comparing bottled versus tap water systems
- analyzing indoor environmental exposure pathways
- reviewing scientific articles for balanced communication
- discussing sustainability versus performance tradeoffs
- evaluating infrastructure and recycling challenges
- or exploring how consumer behavior influences larger systems over time

Students may also benefit from comparing:

- scientific findings
- public perception
- industry realities
- and government policy discussions

A Balanced Educational Approach

One of the central themes of this book is balance.

Several things can be true at the same time:

- microplastics are widespread
- exposure exists
- scientific uncertainty remains
- and practical awareness still makes sense

Educational discussions are strongest when they avoid:

- oversimplification
- alarm-driven conclusions
- and false certainty

Balanced discussion encourages deeper understanding and more thoughtful long-term thinking.

A System Perspective

Microplastics are ultimately connected to broader systems involving:

- manufacturing
- packaging
- transportation
- infrastructure
- consumer expectations
- and environmental management

Understanding those connections helps students move beyond isolated examples and develop a more complete view of how modern material systems operate.

That broader perspective is increasingly important as sustainability, packaging, infrastructure, and environmental discussions continue evolving globally.

About the Author

Brian D. Schell spent much of his career in manufacturing, packaging, and thermoforming operations, where he worked closely with the realities of product design, materials, production systems, transportation, and large-scale industrial operations.

Over several decades, he observed how modern material systems evolved to prioritize durability, efficiency, cost reduction, product protection, and large-scale distribution performance. Those same systems now sit at the center of growing discussions involving sustainability, packaging, recycling, and microplastics.

Brian is also the author of *Producer Under Pressure*, a systems-focused examination of how extended producer responsibility (EPR), packaging regulation, and product design pressures are reshaping manufacturing and consumer industries.

His earlier memoir, *The Buckell Years*, documented the growth of a family-owned manufacturing business and the operational realities behind decades of American industrial production.

These experiences shaped the perspective behind this book.

Rather than approaching microplastics only as a scientific or environmental topic, Brian views the issue as part of a larger systems challenge involving materials, infrastructure, transportation, manufacturing, packaging, consumer behavior, and long-term product design decisions.

This book was written to help readers better understand those systems in practical, plain language.

The goal is not to create fear or oversimplify a complex issue, but to encourage balanced understanding, thoughtful discussion, and practical awareness as scientific research and public understanding continue to evolve.

Brian continues to write and speak about packaging systems, product design, sustainability, infrastructure, and long-term industrial change.

For additional information, publications, or speaking inquiries:
bschell1959@gmail.com
www.schell-ip.com

Acknowledgements

This book was written during a time when public awareness of microplastics, packaging systems, environmental exposure, and long-term material impacts continues to expand rapidly.

I would first like to acknowledge the many researchers, scientists, engineers, healthcare professionals, educators, and environmental specialists whose work continues to improve understanding of these complex issues. The topic of microplastics crosses many disciplines, and much of what is being learned today comes from people working in very different fields working toward a common goal of better understanding how modern material systems interact with people and the environment over time.

I would also like to acknowledge the many individuals throughout my manufacturing and packaging career who helped shape my understanding of materials, production systems, transportation, infrastructure, and long-term industrial change. Decades spent working within real-world manufacturing operations provided practical insight into how modern packaging and material systems evolved and why these systems became so deeply integrated into everyday life.

I am especially grateful for the professional relationships and industry friendships developed over the years through customers, suppliers, competitors, equipment manufacturers, and fellow manufacturing professionals. Many of these individuals openly shared ideas, operational experiences, and practical knowledge that helped broaden my understanding of both manufacturing systems and the challenges facing the packaging industry.

I would also like to recognize the positive influence of the Society of the Plastics Industry and the Thermoforming Division, where participation in industry meetings, plant tours, safety initiatives, and professional collaboration provided valuable learning opportunities throughout my career. Exposure to different manufacturing operations and industry perspectives helped strengthen my understanding of operational improvement, safety systems, material performance, and long-term industrial change.

Special appreciation is extended to the educators, readers, and industry professionals who encouraged the development of this book as a practical and balanced resource focused on understanding systems rather than promoting fear or oversimplified conclusions.

I would also like to thank those who continue working to improve packaging systems, recycling infrastructure, water systems, environmental monitoring, and material innovation. Long-term improvement rarely happens through one single breakthrough. It usually occurs through many smaller efforts repeated over time across industries, institutions, and communities.

Most importantly, I would like to thank my family for their continued encouragement, patience, and support throughout the writing process.

My hope is that this book encourages thoughtful discussion, balanced awareness, and a deeper understanding of the systems surrounding modern materials and everyday life.

References and Suggested Reading

Research and public understanding related to microplastics continue to evolve rapidly.

Readers interested in learning more are encouraged to review current scientific literature, government resources, public health organizations, environmental research, and materials-related publications.

Suggested areas for continued study include:

- environmental science
- toxicology
- public health
- materials engineering
- packaging systems
- sustainability and recycling infrastructure
- drinking water systems
- indoor air quality
- nanoplastics research

Useful sources for additional information include:

United States Environmental Protection Agency (EPA)

World Health Organization (WHO)

National Institutes of Health (NIH)

Centers for Disease Control and Prevention (CDC)

National Academies of Sciences, Engineering, and Medicine

American Chemical Society (ACS)

ScienceDirect

Sustainable Packaging Coalition

Association of Plastic Recyclers (APR)

Circular Action Alliance

Because scientific understanding continues evolving, readers are encouraged to seek current peer-reviewed publications and balanced sources when evaluating new information related to microplastics and environmental health.

Evaluating Information Critically

Because public interest in microplastics is growing quickly, readers will encounter information from many different sources.

Some information may be evidence-based and balanced. Some may be preliminary but useful. Other information may be speculative, exaggerated, or heavily simplified for headlines and social media.

When reviewing information, it may be helpful to ask:

- Is the source credible?
- Does the discussion distinguish between detection and proven harm?
- Are uncertainties acknowledged?
- Does the discussion feel balanced or emotionally driven?

Critical thinking is especially important in emerging scientific topics where understanding continues to evolve.

A System Perspective

No single source fully explains microplastics.

Understanding the issue requires looking across science, manufacturing, packaging, infrastructure, transportation, consumer behavior, and environmental systems.

This broader perspective encourages more informed discussion, more balanced decision-making, and more practical long-term thinking.

Appendix

Questions Consumers Should Ask About Headlines

Microplastics are now discussed regularly through:

- news reports
- social media
- documentaries
- podcasts
- online videos
- and scientific studies

Some information is balanced and evidence-based.
Some is speculative, emotional, or designed primarily to attract attention.

This does not mean concerns should be ignored.
It does mean consumers benefit from learning how to evaluate information carefully.

Complex scientific topics are often simplified into short headlines or dramatic claims. Understanding how to ask better questions helps create a more balanced perspective.

1. Was the Study Done in Humans, Animals, or Laboratory Models?

Not all studies measure the same thing.

Some research involves:

- human observational studies
- animal studies
- laboratory cell studies
- environmental sampling

- or computer modeling

Each type of research provides useful information, but they do not carry the same meaning.

For example:

- a laboratory study may show how particles behave under controlled conditions
- an animal study may explore possible biological effects
- a human study may examine real-world exposure patterns

Early-stage studies often help identify areas for further investigation rather than provide final conclusions.

Understanding the type of study helps place findings into context.

2. Does the Headline Distinguish Between Detection and Proven Harm?

This is one of the most important questions consumers can ask.

Scientists have now detected microplastics in:

- water
- food
- air
- dust
- wildlife
- and human samples

Detection confirms exposure exists.

It does not automatically prove:

- disease
- injury

- long-term harm
- or specific health outcomes

Scientific research is still working to determine:

- what exposure levels matter
- which particle types are most important
- and what long-term effects may exist

Balanced reporting should clearly distinguish between:

- detection

and

- proven health impact

3. Were Exposure Levels Realistic?

Some studies use exposure levels far higher than what most people encounter in normal daily life.

This is common in scientific research because researchers sometimes:

- accelerate testing
- isolate variables
- or examine worst-case conditions

That can help scientists study potential mechanisms more quickly.

At the same time, consumers should ask:

- Were the conditions realistic?
- Would normal daily exposure resemble the study environment?
- Was the dose unusually high?

Understanding this difference helps avoid unnecessary alarm.

4. Was Heat, Friction, or Material Stress Involved?

Material behavior often changes under:

- heat
- repeated friction
- pressure
- sunlight
- or long-term wear

For example:

- microwaving containers
- leaving bottled water in hot vehicles
- repeated reheating
- or heavily worn materials

may create different conditions than normal room-temperature use.

When reading headlines or studies, it is useful to ask:

- Were materials exposed to unusually high temperatures?
- Were products heavily worn or damaged?
- Did testing involve repeated stress conditions?

Material behavior often depends heavily on use conditions.

5. Was the Research Peer Reviewed?

Peer review means other qualified researchers evaluated the work before publication.

This process does not guarantee perfection, but it helps improve:

- scientific quality
- methodology

- consistency
- and credibility

Consumers should recognize the difference between:

- peer-reviewed research
- preliminary findings
- opinion articles
- social media claims
- and promotional content

Headlines often spread faster than fully reviewed science.

6. Does the Article Acknowledge Uncertainty?

Balanced scientific communication usually includes uncertainty.

Researchers often explain:

- what is known
- what is still unclear
- what limitations exist
- and what additional research is needed

Overly confident headlines may ignore this complexity.

Science often develops gradually through:

- repeated testing
- comparison of findings
- improved measurement
- and long-term observation

Uncertainty is not weakness. It is part of how science progresses.

7. Is the Information Emotionally Driven?

Some headlines are designed primarily to create:

- fear
- outrage
- shock
- or anxiety

Emotional communication spreads quickly online because it captures attention.

That does not mean the issue is unimportant. It means consumers should pause before assuming the most alarming interpretation is automatically correct.

Balanced information usually:

- explains context
- avoids absolute conclusions
- acknowledges tradeoffs
- and separates evidence from speculation

8. Does the Discussion Consider the Larger System?

Microplastics are not caused by one product or one behavior alone.

They are connected to larger systems involving:

- packaging
- textiles
- transportation
- infrastructure
- manufacturing
- consumer behavior

- and environmental management

Oversimplified headlines sometimes focus on one isolated product while ignoring the broader system.

Systems thinking creates a more accurate understanding than single-cause explanations.

A Practical Perspective

Consumers do not need to become scientists to think critically about microplastics research.

A few thoughtful questions can help separate:

- balanced information

from

- exaggerated claims

Several things can be true at the same time:

- exposure exists
- important questions remain unanswered
- scientific research is evolving
- and practical awareness still makes sense

Balanced understanding is usually more useful than either:

- panic

or

- complete dismissal

The goal is not fear.

The goal is informed perspective.

www.ingramcontent.com/pod-product-compliance
Lightning Source LLC
LaVergne TN
LVHW022044200926
848481LV00003B/780
* 9 7 9 8 9 9 5 6 8 7 3 4 4 *